HE WHO HAS EARS TO HEAR

THE GALCOM STORY

Gary Nelson

HE WHO HAS EARS TO HEAR, THE GALCOM STORY

ISBN 978-0-9748250-0-7

Copyright © 2004 by Gary Nelson
Cover art work by John Russell
Chapter 16 gives a 2009 ministry update

Published by: Galcom Publications
                11621 Carrollwood Drive
                Tampa, FL, 33618
                Ph. 813-933-8111
                Email: Galcomusa@galcom.org
                Web site: www.galcomusa,com
                            Or
                    www.galcom.org

All rights reserved under the International Copyright Law. No part of this publication may be reproduced, stored in a retrieval system, or transmitted, in whole or in part, in any form or by any means, electronic, mechanical, photocopying, recording, or otherwise, without the prior express consent of the publisher.

Unless otherwise indicated, all scripture references are from the authorized *King James Version* of the Bible.

Printed in the United States by Morris Publishing
3212 East Highway 30
Kearney, NE 68847
1-800-650-7888

Table of Contents

Foreword by George Otis
Preface

Chapters:
1. Who will give them ears to hear?
2. In the Beginning - The Genesis of Galcom from the pens of the Founders.
3. One Radio.
4. The Mighty One Visits Iraq.
5. The Mongolian Connection.
6. A Harvest in Moldova.
7. Wings of Mercy - Mexico.
8. White unto Harvest - Bolivia.
9. Building in Latin America.
10. A Harvest in South America.
11. Reaching into The Caribbean.
12. God Rains on the Rain Forest.
13. All Over Africa, God's Spirit is Moving.
14. Reaching other Nations.
15. Memories of Travels to Many Lands (by: Allan McGuirl)
16. The Stones are Crying Out, Plus a 2009 radio update.

Acknowledgements of donors

Foreword

My heart burned as I read this book of real testimonies from around the world. It reminded me of the countless stories in the book of Acts and the Gospels. The Bible says, "They turned the world upside down." Through a remarkable breakthrough, Galcom conceived and birthed the compact solar powered radios, which are similarly impacting the nations with the Gospel message.

I remember one Maronite leader who regularly moved through Jordan, Israel and Syria, breathlessly telling me about people who could now hear the Good News of salvation and miracles. He reported that 22 new churches had been birthed as a result of the radio outreach in Syria alone. Through the compact Galcom radios, they suddenly <u>had ears to hear</u>. Prayers to the living God, flowing out of the airwaves, resulted in captives freed, diseases cured and renewed hope. To God be the glory!

Hundreds and hundreds of thousands of these small miracle radios are scattered into every continent to bless the hearers and to inspire each of us to pray for Galcom and to help them to speed these radios to those who want and need them. They are probably covering only one or two percent of those now in need and who want this help. What a soul-winning opportunity this makes for each of us on behalf of advancing His kingdom. Whole communities, villages and cities are ready and needy to receive God's Word as we provide them ears to hear.

I had the privilege, through the Lord's calling, to found a Christian broadcasting network. Its programs fell like golden rain into every continent on this earth. I consider the achievement of Galcom's founders - Harold Kent, Ken Crowell, and Allan McGuirl - to greatly exceed the spiritual harvesting of my own work.

Today is God's soul-harvesting season. The Lord said that the days were coming when the reaper would actually race in front of the sower. Isn't it glorious to witness this beginning to happen through the Go-Ye radios? We want to rush, to speed this along through our prayers and generous offerings.

George Otis, Founder
Kingsworld Ministries

Preface

"How then shall they call on Him in whom they have not believed? And How shall they believe in Him of whom they have not heard? And how shall they hear without a preacher?" **Romans 10:14**

There are still billions of unreached people around the world who have never heard the life saving message of Jesus Christ, <u>and can't read</u>.

This book: **HE WHO HAS EARS TO HEAR,** is the success story, in testimony form, of what God has done through Galcom International, a ministry that is taking the Gospel, by Christian radio, to those who have never heard.

Galcom International was formed in 1989 by three men and their wives, Harold and Jo Ann Kent, Tampa, Florida; Ken and Margie Crowell, Tiberius, Israel; and Allan and Florrie McGuirl, Hamilton, Ontario, Canada.

God impressed upon each of these three men the need to place radio receivers in the hands of millions of people around the world who can't read, but are within range of a Christian radio station which is broadcasting the good news. These three couples lived in different countries and did not know one another. Through the intervention of Dan Karvonen in Mankato, Minnesota and Paul Maass, in Tampa, Florida, the linkup was made and the three men came together for the first meeting in February of 1989.

At this first meeting each person was surprised that the others had the same vision. Realizing that this project was guidance and direction from God they founded Galcom International with offices in Tampa, Florida and Hamilton, Ontario, Canada. By the year 2004, 15 years later, Galcom International has distributed over 420,000 Go-Ye fix-tuned solar powered radios in 118 countries and assisted in the installation of dozens of low-powered Christian radio stations on remote mission fields. These radio stations and the radios are now taking the Good News of Jesus Christ to millions of people around the globe on a daily basis.

This book is a compilation of testimonies from some of these countries plus a chapter <u>on The Genesis of Galcom</u>, how it all began, in the words of the founders.

In the Bible, in 2 Sam. 23 and 1 Chro. 12, King David's mighty men of valor are described. I consider the founders of Galcom International to also be <u>mighty men of valor</u>.

A businessman, Mr. Blades, recently stayed in a Holiday Inn in Shreveport, LA and saw a plaque on the lobby wall with the following words which he relayed to us.

Je' Raphe defines a man of valor in the following terms:
> **There comes a time when a man**
> **Steps into the purpose,**
> **Anointing and empowerment of God.**
> **He may be a brother, a son**
> **A spouse, a father, or a friend.**
> **But most of all he is a Man of Valor,**
> **Not a man that fits any particular mold**

But instead, having a will and a spirit
That encourages others through actions,
Gifted, talented, knowledgeable,
Offering insight, but not dictatorship,
You experience his strength and love.

He is a man that embraces the Word of God,
A man that prays, listens,
And is obedient to the voice of God,
Having the ability to give of himself
In order that another may be empowered.

He is a standard setter in every aspect.
Mighty, yet gentle, and sensitive;
A man of substance - a mentor of others,
Prestigious by the power of God
Steadfast in a Life for God
A Man of Valor.

I can see why God selected the three men listed above to start this worldwide ministry, taking the Gospel to the ends of the earth. They are men of valor, well described here. Yet, the stories contained in this book are not what these men have done, but are testimonies of what God has done as men have obeyed God and allowed Him to work.

Gary G. Nelson
President/Galcom International USA, Inc.

Chapter One

Who Will Give Them Ears to Hear?

"Behold, I stand at the door, and knock; if any man hear My voice, and open the door, I will come in to him and will sup with him, and he with Me. . . He that has an ear, let him hear . . .." Rev. 3:20, 22

Thousands of Native Americans from Mexico now live high in the mountains of the Sierra Madre Occidental Range of Central Mexico. The Spanish conquest in 1520 under Hernando Cortes drove these people out of the flatlands into these high mountains. They fled to avoid enslavement in the mines and on plantations. They still live in these rugged mountains where some canyons are 11,000 feet deep. Thousands of huts line these canyons and mountains where the people try to eke out a living as best they can.

In one of these huts lives a 62 year old woman who had been suffering with sickness for 40 years. She had been visiting the tribal witch doctors without any sign of improvement. Her life was one of suffering and boredom. In this remote area she had no electricity, no radio or TV, and no telephone. She could not read as she had never been to school.

One day while sitting in her hut she heard the strange sound of an aircraft engine directly over her hut. She went outside and looked up to see a small parachute with a box attached coming down close by the front of her hut. She wondered, "Could it be an explosive device or something harmful?"

After putting aside her fears she decided to go get it, open the small box and inspect the contents. Inside she found a small red plastic device plus a pictorial set of instructions showing her how to operate the device (a Galcom solar-powered SW radio). Following the instructions she turned the radio on and began hearing Christian broadcasting, proclaiming the Gospel of Jesus Christ, for the first time in her life. God had come knocking at her door and the Galcom Go-Ye radio had given her ears to hear.

The radio signal was coming from KVOH Voice Of Hope Christian radio, broadcasting on short wave in Spanish. For days she enjoyed the Christian music, Bible reading, and Gospel messages. Soon she prayed the salvation prayer following along on the radio and gave her life to the Lord Jesus Christ. A few days later, while listening to the program, she was told that Jesus could heal and if she would lay one hand on the radio and the other on her body and pray along with the preacher on the radio she could be healed. As she prayed the prayer of faith she received instant healing.

She became so excited. She thought, " If Jesus can save me and heal my body, He can teach me to read the book called the Bible that they keep talking about on the radio." She went on a quest to find a Bible in Spanish. After a few days she acquired a Bible and asked the Lord to teach her to read it. Within nine days she was reading the Bible as she followed along with the radio Bible readings.

This was too good to be kept to herself. This Jesus, to whom she now belonged and served, had saved her, healed her, and taught her to read His Word. She felt that she must share this with her relatives and friends living in this rugged mountain canyon. She began inviting them over for Bible study and prayer. Soon she had a small group meeting regularly; the first "House Church" in her valley.

The amazing facts of this story are that this woman was saved, healed, learned to read the Bible and started a home Bible study without ever having met another Christian. No pastor, evangelist, or missionary had ever been up in those rugged mountains to share the Gospel with this woman. She had just received a Galcom radio dropped out of the sky right to her mountain front door in an operation called "Wings of Mercy".

Missionary pilots Jerry Witt, Jerry Wiley and Alex Fedorenko have been delivering these precious packages for several years to the people of Mexico living in the high, remote mountains and who could not be reached with the Gospel in any other way.

Galcom International is a unique ministry that is using the tools made available by modern technology to take the Gospel message to millions of people who are illiterate and difficult to reach. Using Christian radio as a primary tool Galcom has touched unreached people in 118 countries around the world.

If there is a Christian station broadcasting in the area, Galcom provides high quality, solar-powered radio receivers (called Go-Ye radios) that are fix-tuned only to that station. If there is not a Christian radio station in the area, Galcom

International will work with a local church or mission agency working in the target area and help them install a low-powered Christian station. Then Galcom supplies Go-Ye radio receivers tuned only to that station to be distributed to potential listeners who have never heard the good news, can't read or do not have electricity.

Many times people in these mission areas are so poor that they can not even afford batteries for a radio. Our generous God supplies solar power free of charge to everyone. Galcom's solar-powered radios equipped with rechargeable batteries will operate for years in these remote areas of the world.

The following chapters tell some of the Galcom story, which is a story of what God is doing to reach the lost, using men and women who are obeying the call of the great commission.

Chapter Two

In The Beginning; The Genesis of Galcom from the pens of the Founders

"These are the names of the mighty men whom David had;" 2 Samuel 23:8a

The Lord raises up mighty men from each generation, called by Him to carry on His work. In 1988 God spoke to three men, calling them to do a work for Him. He gave them an innovative way to reach the lost in nations where the people are illiterate, have no modern conveniences such as electricity, and have very little contact with the outside world.

Although these three men were living in three different countries and did not know one another, God gave them each a common vision of how this work was to be done. They each understood the times, knowing what to do, as King David's mighty men, the men of Issachar, did as described in 1 Chro. 12:32. By supernatural "coincidence" God brought these three men together in a bond of friendship and common purpose and Galcom International was born.

The following are their stories, how God spoke to each of them and how they obeyed the voice of the Master, the voice of The Lord of the Harvest.

**From the pen of Harold Kent
Founder of Galcom International
Bright Lights for the Future**

In the fall of 1988, as I was sitting in a conference totally consumed in my desire to serve God, He spoke to me and I heard my Father say, "Make fix-tuned radios and distribute them to people around the world who have no way of hearing the Gospel."

In 1965, at the wonderful age of forty when, it is said, life begins, I was born again into the Kingdom of God. After seeing my wife JoAnn give her life to the Lord and then living for Him, I finally said, "Yes." I can say my life made a slight change after that, but it was in 1968 that a major change took place.

I was approached with the idea of taking a tour to Israel. I didn't think I wanted to go, so I put a fleece before the Lord saying that He would have to show me if I were to go. One day in my office, I picked up a magazine about growing vegetables. I had been wondering what would be the best way to water my orange trees and I saw that a system for watering vegetables was featured in this magazine. As I read the article I got excited. I thought, "This would work great for trees too." Quickly I looked at the end of the article. The place to find out about this drip irrigation system was in Beersheba, Israel. I realized this was the answer to my fleece before the Lord. We were to go to Israel.

So, in 1968, my wife and I found ourselves on a plane bound for Israel with Derek Prince and his tour group. As with most tours to Israel, the time came when a baptism in the Jordan River would take place. The night before the baptism Derek spoke on the subject. I did not plan to be baptized but as we approached the Jordan, a surprising change began to take place in me. I found myself joining the others in the Jordan River. Thank the Lord I did as my life has not been the same since. I explain it by saying, "My life changed 180 degrees." I was headed for a mediocre life but when I was baptized my life became meaningful and full of the love, grace and the mercy of God. Life became exciting.

Shortly after making Jesus Lord of my life, two years before our trip to Israel, my family and I were attending Christian camp. As we were sitting in an assembly of Christians, listening to an anointed message, the Lord gave me a vision. I saw the world as a dark globe. It was a beautiful deep blue. As I watched, I saw a dot of light appear, then another dot, then another until the whole earth was filled with dots of light. I did not know the meaning of this vision but this picture stayed in my mind. It was many years before I would understand the vision.

My wife, JoAnn, and I have had, for many years, a missions vision to reach the unreached of the nations with the Gospel of Jesus Christ. In 1988, we attended a conference held by Derek Prince Ministries in Ft. Lauderdale, Florida. For many years we had been regular listeners of Derek's fifteen minute radio program which was broadcast nationwide at that time. At this conference Derek spoke of desiring to broadcast radio programs on short wave radio so that the Gospel could be preached to those people who lived in unreached parts of the world. The thought crossed my mind, "How do you know if anyone is listening?" In many parts of the world, the people do not have electricity or access to a radio. Even the ones that might have a short wave receiver must by chance tune into the program. It was at that moment I heard my

Lord say "Make fix-tuned radios and distribute them to people around the world who have no other way of hearing the Gospel." I realized that this Word from God was confirming the vision that I had previously, and it would affect the whole world, allowing people to hear the Gospel on the high powered radio waves.

It is one thing to be given direction to do something, but it is entirely something else to make it come about. I wondered how to find such a fix-tuned radio or if I should have one made. Not knowing much about radio, I did not know where to get them made. The problem that I was now presented with was how to get such a radio designed and produced in the quantities needed and then distributed around the world. I began to pray that God would show me how to do this. I thought about going to China and getting radios produced there. I tried to make some contacts in Hong Kong but to no avail. It was a couple of months later that I would have a major breakthrough.

One day, a friend from Minnesota, Dan Karvonen, was in Tampa on business. He called to invite me to have lunch with him and a gentleman from Israel, Ken Crowell, whom I had not previously met. I had told Dan about my idea of a fix-tuned radio project. At lunch Dan shared some ideas about the radio project and told Ken that I was seeking an answer as to what to do with my vision of making fix-tuned radios. Ken was CEO of Galtronics Electronics located in Tiberias, Israel. As I talked with Ken about my vision for fix-tuned radios, his mouth literally fell open and he became very excited. He related to us that he had been given a similar vision of producing fix-tuned radios some time before. Ken excitedly told us that he was an electronics engineer and had spent much time engineering such a radio but had not carried it further as he did not know what to do next.

He had placed those plans in an envelope in his desk drawer at home. I knew our contact had to be the leading of the Lord.

The two of us determined to work together on the project. Ken had heard of a man in Canada by the name of Allan McGuirl who also had the same vision and was working with Gospel Recordings in Ontario. Allan was acquainted with many missionaries all around the world. He also was acquainted with many mission groups doing Christian radio broadcasting, so was well prepared to distribute the radios to the unreached.

We all arranged to attend the National Religious Broadcasters (NRB) convention in Washington DC and to meet there. In our meeting we decided to work this calling together. Ken would finish engineering the radio, I would begin funding the project and Allan would distribute the radios around the world. We would be a threefold cord not easily broken. On August 15, 1989, GALCOM INTERNATIONAL WAS BORN.

We started radio production in Tiberias, Israel and our first 40,000 fix-tuned radios were tuned to Voice of Hope, AM 945 broadcasting from Lebanon, and were sent into Lebanon, Jordan and Syria. Testimonies began to arrive and with that encouragement, production went into full swing. As of this writing, 15 years from that starting point, we have sent out over 420,000 Galcom fix-tuned radios to over 118 countries. They have gone to the most needy parts of the world making the Gospel message available to millions of people who would not have heard the saving message of Jesus Christ in any other way. Most of these radios are solar-powered and fix-tuned only to short wave or local Christian radio stations.

The vision I had way back in 1965, a vision of the world going from darkness to light, was now coming true as the light of the Gospel is beginning to flood the nations. I can say today that my heart rejoices as I see the fulfillment of that vision. We receive testimonies from many people throughout the world going from spiritual darkness to spiritual light. I can only say that I am overwhelmed with what God has done. Thank you Lord, for Your faithfulness.

From the pen of Ken Crowell
Founder of Galcom International
The Galcom Story

I remember the day that Galcom began in the lives of Margie and me over 15 years ago. I was sitting at my desk in my small office at Galtronics in Tiberias, Israel. Outside, in our little town located at the edge of the Sea of Galilee, the people were full of stories about Christian "missionaries" who were starting a company in their city. The Ultra Orthodox were doing a thorough job of spreading fear and condemnation as to the purpose of Galtronics, and we had become the target of the militant anti-missionary movement in the north of Israel.

Galtronics is a tentmaking, Kingdom business established in Israel in 1978. The company designs and manufactures wireless communication products. Its core technology is related to small antennas, and Galtronics currently produces over 40 million antennas a year for the worldwide cellular industry.

The company was started with three distinct goals:

1. To locate a factory in an area where there is need for a Christian witness, and to help establish a local church.

2. To give employment to believers and non-believers, creating an environment of trust and knowledge of Jesus as Messiah.

3. To bless the Nation of Israel by giving jobs and exporting products.

Today, 25 years later, Galtronics is well respected as the largest employer in our area. The Israeli Government has awarded the company with the highest honor in the nation for blessing and contributing to the growth of Israel. Only God could have done this! Today we are seen as a blessing to the Nation of Israel, but this was not so in the beginning.

Back to that day in 1988, as I sat at my desk, I reached in my file and took out a dusty manila folder. "Lord," I said, "I have been disobedient. One year ago you asked me to design a fixed-tuned radio. We did as you asked, but I forgot about the project and put this file in the back of my cabinet. Lord, you know we have enough trouble from the Orthodox without building a product to be used in missions, but if this is what you want, I am ready to move ahead according to your will."

A few days later, a friend of mine, Dan Karvonen, phoned and asked if I could come to Florida and meet a person who has a proposal he felt might interest me. It just happened that I was to fly to the USA that following week on business. This is an example of the Lord's leading if we will just listen.

The next week I found myself in a hotel lobby in Tampa, Florida, being introduced to a new friend, Harold Kent. Harold said he had a proposal for me. If I would be interested in designing and building a fixed-tuned radio in Israel for use in worldwide Christian broadcasting, then he would provide the funding. How could he have known about that manila folder in my office in Tiberias?! This had to be just another chapter in God's plan for both Harold and me.

Upon returning to Israel, I asked our Galtronics engineers to prepare for radio production. A place was set in our manufacturing facility and Galcom's Go-Ye radios became a reality.

About the same time I heard of another person, Allan McGuirl, who was interested in the fix-tuned radio ministry. He was located in Canada and was working on a design at home in his basement. Allan was working for Gospel Recordings as a full time missionary. He was doing a fine job, but God had set his mind on the radio ministry.

I phoned Allan and told him about Harold's and my interest in the radio ministry and we set a plan for Margie and I to visit him in Hamilton, Canada. During our time together it became evident that God would have us work together. Harold would do the funding, Galtronics would do the manufacturing, and Allan would do the distribution. We all agreed, and the new ministry was born.

I would like to acknowledge some others who have been instrumental in making the vision come to pass. God brought together a team of people needed to get the job done. I have already mentioned Dan Karvonen, who first arranged the meeting to bring the team together and has been a valuable member of the board since inception. Paul and Jeanette Maass did the initial administration of Galcom in the USA and have also been continuous corporate directors. In 1991 Gary and Mary Nelson picked up the ball on donor relations and other administrative duties. Noah Garaway got the production of radios off the ground in the plant in Israel. God raised up the right people for the job who came from various walks of life.

The name chosen for the ministry was Galcom International. Psalm 37:5 says: "Commit thy way unto the Lord; trust also in Him; and He shall bring it to pass".

Commit in Hebrew is gal. Galgal in Hebrew is a wheel. Gal is one half of a wheel, or a wave. A wave goes on and on. Gal means commitment or rolling our needs and cares over to God daily.
Com is an abbreviation for communications. International means worldwide.
So Galcom International indicates our commitment to communicating the Gospel worldwide.

We decided to call our little solar powered fix-tuned radios, the Go-Ye radio. This term is taken from the great commission in Matthew 28:19, "Go ye therefore, and teach all nations. . . " These little talking missionaries are now in many parts of the world, 118 nations, tuned to Christian radio stations, preaching and teaching the Word of God to people who would not have other means to hear about Jesus. Go-Ye radios give them ears to hear.

**From the pen of Allan McGuirl
Founder of Galcom International
How God directed me into Galcom**

As I traveled in various parts of East Africa in remote communities, I praised God for the many who were hearing the Gospel in their own language through the use of little cardboard record players and hand wound cassette players. I was, by God's grace, the Canadian Director for Gospel Recordings, but my heart was stirred as I had listened to reports months earlier of how a group of radio leaders from HCJB, TWR, FEBC and ELWA had united hearts to reach millions of unreached people by radio by the year 2000. This "World by 2000" project started me thinking. Many of these people in remote villages have no radios, no electricity and little money to buy batteries. I sensed that God was impressing upon me to build a solar-powered fix-tuned radio.

I had mentioned this a number of times to our mission's leadership, but their hearts were fixed on the recording aspect of unreached languages. During the summer of 1988 I built a prototype of a solar fix-tuned radio. It still sits on my office shelf as a reminder of God's direction in this area. In September, 1988, at our Gospel Recordings board meeting, I presented this first working model of an AM solar fix-tuned radio tuned to CFRB 1010 in Toronto. The sound came through clearly with lots of volume. That was as far as it went at this time.

My heart was really burdened as I wrestled with this concept of making fix-tuned radios for the mission field. "Lord, we need your guidance," I prayed. A short time later, I received a call from Ken Crowell in Israel. Apparently, he had heard what I was doing with a fix-tuned radio. Ken had already been to Tampa, Florida, and Dan Karvonen had introduced him to Harold Kent. Both Harold and Ken had now been challenged to build a fix-tuned radio under their own interesting circumstances.

In September of 1988, Ken and his wife Margie flew to Canada and met me at the Gospel Recordings office on Aberdeen Ave. in Hamilton, Ontario. I was just getting ready for an IFMA Conference. As we discussed the concept of fix-tuned radio, I demonstrated the model I had made. There was a strong interest in all of us joining together to make radios.

Our next meeting was in early February, 1989, at the National Religious Broadcasters Conference. I recall one seminar with Dr. Ron Cline of HCJB and how people were excited about the concept of a little fix-tuned radio. We left elated at the interest but not sure how God was going to put it all together. The basic concept was that Ken could make the radios in Israel, Harold would sponsor funds to get things started and I would work on determining needs and arranging distribution.

After the conference, I sent some names of organizations for Ken to contact from Israel and see what interest there might be. When responses started to return to Israel, the ultra Orthodox caused much disruption in the flow of mail. It was evident that correspondence and contacts would have to be handled elsewhere. Ken contacted me in early May to consider making a change in ministry and locating the Galcom office in Canada.

My wife, Florrie, and I were much in prayer. Would our supporters continue with us in this new line of ministry? With five children, this was a major concern. We didn't have a name, a finished product, registration with the government, or even an office. In June, 1989, the Lord showed me 1 Chronicles 4:10, the prayer of Jabez, and a real peace flooded my heart. Florrie and I felt this was the direction God was leading us. We informed Ken of our decision. By God's grace all of our supporters but one stayed with us and we left Gospel Recordings and started Galcom International in Canada.

In early August of 1989, Florrie and I flew to Israel where we met with Ken and Margie Crowell and Noah and Gila Garaway. What a precious time as we waited upon the Lord to guide us. Ken came up with the name Galcom which means "A Commitment to Communication". I suggested the word International as we could possibly get into other countries outside of the Middle East. Florrie formulated the purpose "To provide durable technical equipment for communicating the Gospel." Later "worldwide" was added.

Back in Hamilton we pushed our dining room table aside and set up an office. On Monday, August 15, 1989 at 9:00 a.m., Florrie and I stood in our dining room and, as Psalm 37:5 recommends, committed the work to the Lord. The basement workbench became our R & D department. Ken set up production to build radios in Tiberias and Harold funded the fledgling ministry. With the contacts that I already had, I began praying for the Lord to open doors.

Little did we realize that, 15 years later, Galcom would be a household word among many mission organizations, over 420,000 solar fix-tuned radios would be distributed in 118 countries, hundreds of thousands of people from around the world would be hearing the Gospel on Go-Ye radios, many thousands would respond to the claims of Christ, churches would be formed, witch doctors would be saved and many ministries would be enabled to reach out farther than ever before because of these mini-missionaries. Truly this has been God's doing and we give Him all the glory.

Chapter Three

One Radio

"Behold, I send an Angel before you to keep you in the way and bring you into the place which I have prepared." Exodus 23:20

Occasionally we will hear a story of how just one radio has reached many people for Christ. It is as if God has sent His angel before each radio and prepared the way. The following are some stories of what one radio can do in the hands of a Mighty God:

One radio in Haiti
Men for Missions International has been distributing Galcom radios in Haiti fix-tuned to 4VEH Christian radio in Cape Haitian as part of Operation Saturation Haiti. During a recent trip to Haiti they decided to follow the steps of just one radio to access the results of the Galcom Go-Ye radio outreach. Here is what was reported by Gene Bertolet, OpSat Prayer Coordinator:

"One of the outstanding events during a recent trip to Haiti by a Men for Missions group was to follow the steps of one radio which had been given to an EE technician, who in turn had given the solar powered radio to her father. After listening to Radio 4VEH for about three months, her father accepted Jesus as his Savior. He then gave the radio to his friend, a witch doctor. The witch doctor, too, accepted the Lord.

"One morning this young EE technician took our team to the home of her father, about 45 minutes away. As the father gave his clear testimony others gathered around. After two hours 17 more people had accepted the Lord. Then they went to the witch doctor's home, which was a fifteen minute walk through the countryside into a valley. As they came up to his home, there was a white cloth covered table with a Galcom solar-powered radio in the middle of it.

"When the witch doctor came out of the house, he graciously seated his guests. Once again they heard a clear testimony of salvation. He said, 'No one comes to me as a witch doctor anymore, for they know I have accepted Jesus.' Twelve of his thirteen children had accepted Christ, but not his wife. As people gathered around, one of our team walked over to the witch doctor's wife who was shelling corn. In a caring manner he talked to her and others nearby. At the close of their time together, five more had accepted the Lord, including the witch doctor's wife.

"One of the team members said, 'As I think about Operation Saturation - from this one radio, 36 people accepted the Lord. What an investment.' And this all happened within the space of four months. " Gene Bartolet

Operation Saturation has already distributed over 15,000 Galcom Go-Ye radios in Haiti in the last two years. Galcom has produced and distributed throughout the world over 420,000 fix-tuned radios in the last 15 years. This story from Haiti shows the outreach of just one radio and also the potential outreach of 420,000 Go-Ye radios in the hands of the Lord. Our God is an awesome God and He can take something small and make it bear results.

One radio in Ecuador

We recently heard about a village in Ecuador where 98% of the people were saved through one

Galcom radio. A missionary pastor in Ecuador, Felix Lucian, related to us the following story:

"A few weeks ago one of the pastors in the Association of Interdemominational churches here in Ecuador, returned from a 10 day mission trip with three young people. They found a village at the mouth of a river where they intended to do mission work, both in the village and up river. What they found was that 98% of this village at the mouth of the river were already Christians. Earlier someone had given them one Galcom fix-tuned radio and they had attached it to an external speaker so that the entire village could listen to the Gospel programming. Also, without prompting from the visiting missionaries, they offered themselves, their homes, and their village as the center of operations for mission trips to evangelize upstream and throughout the district. What impacts us is that they have been and continue to be discipled in Christ by one missionary radio."
Pastor Felix Lucian

One radio again in Ecuador: A report from Good Shepherd Radio station in Ecuador

One morning a local shoemaker was running the station's morning 6:00 a.m. programming as a volunteer. A Quechua Indian, with his wife and family came into the station. The Indian man explained that they had come a great distance from a remote village. This village had only one "Go-Ye" radio which they hung in a tree so that the entire village could gather and listen to the Gospel broadcasts in the Quechua language. He and his family had been listening to the program for some time and he took up an offering for the radio station. He came to deliver the offering to the station which amounted to a little less than $1.00 US. He and his family walked 10 hours to a bus stop then rode the bus for four more hours. After he presented the offering to the station he asked how he and his family could come to know Jesus. The station operator put on a long play album for the morning broadcast and then went outside with the Indian family and led the entire family to the Lord. That family left full of joy and the new gift of eternal life.

One radio in Panama

Recently a Christian missionary was traveling on a remote trail in a jungle area of Panama. He was stopped by a local villager and when the villager found out he was a Christian he asked the missionary if he would come into the village and serve communion to the Christians of that community. The Christian worker asked if any other missionary, pastor or evangelist had been to the village. The man said "No but there are a large number of Christians in the village." The Christian worker asked how they had heard about Jesus, and the man pulled out a radio tuned to the HCJB station there. He said they had been saved and started a village church listening to this little radio and now they wanted someone to come and serve them communion. It seems that an angel of The Lord had gone before this missionary and prepared the way.

One Radio in Puerto Rico

Janet Lattrell has a prison ministry to as many as 6000 prisoners in Puerto Rico. One of the prisoners named "Big John" was very hostile toward everyone but she was able to get him to accept a Galcom radio. He later testified that as he listened to Christian radio he learned how his burdens could be lifted if he received the Lord. He did repent and accept Christ then asked Janet to start a Bible study in his cell block.

He said that the very day she gave him the radio he was going to commit suicide. The Gospel through this one radio saved his life, his soul, and lifted his burdens. Now there is a Bible study in "Big John's" cell block with many souls being brought into the Kingdom because of one Galcom Go-Ye radio.

One radio in Bolivia

Bob and Beth White moved to Cochabamba, Bolivia in May of 2002 to be a part of the ministry there to reach the 2.2 million Quechua Indians living throughout that country. They have established Mosoj Chaski Christian radio broadcasting in the Quechua language to the entire country. The following is a testimony of the results of one radio given to a Quechua man:

"A little while back some folks were out in the country giving out Galcom solar powered radios which only pick up the Quechua radio station from Cochabamba. Some men from a village some distance away asked what their ministry was all about, so the team gave them one radio and sent them home. Later they heard that because of this one radio, there are now 10 families from their village who want to become Christians. The area where these men live is a totally unreached, unchurched area, but now, through one Go-Ye radio, there are 10 families who want to know the Lord. Two evangelists went out to this area (a 4 hour drive) to explain the way of Salvation to these families and to encourage them in their walk with the Lord.

"Please pray especially that satan will be held back and that these families will have total freedom from the fear of bondage that they have had in their animistic religion and that they will come to know Jesus in a real life changing way. Pray for the community where they live, that the other people there will also want to know Jesus. Satan will want to play havoc in that area now, so pray that nothing will happen that could cause the rest of the community to blame Christianity for the things that go wrong. We are so excited to see a new community being won for Christ through Christian radio with only one Go-Ye radio, and are looking forward to what Jesus will do in that area through these new believers." Bob and Beth White

Throughout the years we have heard many stories of entire villages being converted listening to one Go-Ye radio, of churches being planted using one Go-Ye radio, and of many entire family units coming into a relationship with our Lord and Master because of one radio. Our prayer continues to be: "Lord, continue to send your angels before each radio and bring it into the hands which You have prepared."

Chapter Four

The Mighty One visits Iraq

"**But the Lord is with me as a Mighty Terrible One; therefore my persecutors shall stumble, and they shall not prevail; they shall be greatly ashamed; for they shall not prosper; their everlasting confusion shall never be forgotten. . . Sing unto the Lord, praise ye the Lord; for He hath delivered the soul of the poor from the hand of the evildoers.**" Jer. 20:11,13

Iraq's population is composed of 96% Muslim, 3% Christian and 1% other religions. Until recently Christianity has been opposed and restricted socially. With the change in the political regime in Iraq we have great hope for increased opportunity to share the Gospel by expanding radio outreach.

Presently we are working with Servant Group International who is partnering with the National Protestant Evangelical Church in Northern Iraq, where they are involved in operating three FM Christian radio stations in the towns of Dahuk, Sulymaniyah and Arbil. These Christian radio stations, called **The Voice of Peace,** have a potential listening audience of 1,250,000 Kurds. Thousands of Galcom fix-tuned radios have been distributed to the people of the region throughout Northern Iraq within the broadcast range of these radio stations. The Kurds are very receptive to the Gospel, are coming to the Lord and being taught over the radio.

A local pastor there, Yousef, reported that the radio stations are now broadcasting the Good News in Kurdish, eight hours a day. Government officials are requesting Go-Ye radios, ladies are carrying the radios with them into their offices, teachers are listening to the station during breaks, bus drivers play the radios with Christian music for the passengers, shop keepers in the markets are listening to the radios loudly while before they had to listen secretly. People in outlying villages are asking for the broadcast to cover their villages and want radios. Pastor Yousef says, " Exciting isn't it?" We are encouraged by the success of this Gospel outreach among the Kurds of Iraq. God has allowed us to have a small part in what He is doing among these people.

Douglas Layton, International Director of Servant Group, after returning from the region, shared that Galcom radios are now seen and heard everywhere, sending out the Good News. Doug said that he could use 10,000 more Galcom GO-YE radios for Iraq. He reports that there has been a great response to the Christian broadcasts and the Galcom radios.

Servant Group International periodically sends out reports about the dramatic results of radio outreach in Kurdistan, Northern Iraq. Two of their Christian radio broadcasts come out of stations located in Sulymaniyah and Arbil along the Iraq/Iran border so the message is even reaching deep into the Nation of Iran.

Here are some excerpts from a trip report to Northern Iraq by Brandon Verner, International Field Director of Servant Group International:

"The Voice of Peace Christian broadcast continues to be a front-runner ministry for the Kurdish Church. The radio station in Sulymaniyah has developed a special connection with the Classical School of the Medes (CSM). At the official opening of the CSM in Sulymania, we handed out over 200 Galcom fix-tuned radios to the parents of the students. We were even able to give radios to all the officials from the Ministry of Education. The next day, representatives from the Ministry of Education came to the school requesting more radios for their office workers. We were happy to provide them.

"This broadcast is able to penetrate homes and offices and hearts of those who would normally not have any exposure to the Gospel. We have found that radio is one of the best tools for breaking ground and planting seeds in the hearts of those we seek to reach here in Kurdish Iraq. The radio broadcasts represent a unique way for the Kurds to be exposed to the Gospel in a very non-threatening way."

The report goes on to say: "Through these radios we have access into houses and offices without having to knock on the door. Families who would normally be opposed or afraid of having a relationship with a Christian or a church, are hearing peaceful music and loving teaching on these radios that breaks down their false beliefs about Christianity. Radio can befriend such a wide variety of people; girls and boys, men and women, young and old, educated and uneducated. One great advantage is that we are able to reach those that speak both Kurdish and Arabic by offering broadcasts in both languages. We are producing 60 Christian worship songs in these languages. Music is something that easily penetrates cultural barriers.

"Because of the danger involved and recent heightened state of alert in that area, the radio station staff in Sulymaniyah, the city closest to the Iranian border, was thinking of closing down the radio station for awhile. As they prayed about it they received a Word from the Lord from Jeremiah 20:9-13, to continue to proclaim the Lord and He will prevail. He is with us, the Mighty Awesome One."

They continued to broadcast and not to shut down the fire of the Gospel. Now more than ever there is a need for the Prince of Peace to minister to the hearts of the Muslims there.

Two Testimonies from Iraq
1. A young married Muslim woman, 16 years old, accepted the Lord and her husband found out. He had her put in prison with her one-year-old son. The husband did this in the same month that she had lost her mom and only brother in a car accident. The local church stood by her in prayer. In court the judge asked her if she had become a Christian. Trembling with tears in her eyes she asked the judge, "Do I have the right to chose my faith?" The judge told her "yes" but when she then told him she had become a Christian the judge divorced her from her husband at the husband's request. The amazing thing was that the judge allowed her to keep her baby and raise him as she chooses until he becomes 16 years old, at which time he will decide for himself his faith. Many times these people pay a high price for their faith, but the eternal rewards are far greater.

2. A Christian man named Dilshad was sharing the Gospel with people he met. One man responded that he had a dream about a visitor who would come to him and share the same stories that Dilshad was sharing. The dreamer went on to share that in his

dream he was told to listen to a man named Esau (which is the name for Jesus in Arabic). Dilshad then told him that as a child his family called him Esau, but then changed his name. Dilshad then told him about Jesus, the Lord and Savior. The man accepted The Lord and is now in a Bible study with Dilshad.

God is beginning to open more doors for the Gospel to the Muslim world. The Galcom fix-tuned radios are an excellent tool to cross these cultural and religious barriers.

A Martyr in Iraq
Zewar was a Christian living in Zakho, a city in Northwest Iraq. He had a wife and three children and ran a taxi service to support his family. He would give our Galcom radios tuned to the local Servant Group radio station to many of his customers. Toward the end of February 2002, Zewar was having coffee with a customer (which is the cultural norm) before driving him to the requested destination. In the course of the conversation, it was demanded of Zewar that he renounce his faith in Christ. Zewar refused to renounce Christ and the customer drew out an automatic firearm and shot Zewar 28 times. He was shot for his faith and for distributing Galcom radios. There have been no known arrests for this brutal crime. It is our prayer that Zewar will be the last Christian martyr in this country of Iraq.

Since the Iraqi Freedom war of 2002, Servant Group leaders are very excited about the possibility of expanding radio ministries into Kirkuk, Mosul, Basra and Baghdad. The extended broadcasting range will allow millions more Iraqi people to hear the Gospel. Galcom International is now working with

Brandon Verner of Servant Group International on the technical aspects of installing these additional Christian radio stations to reach the people of Iraq. We labor in the harvest field to bring in the full Bride of Christ as we prepare for the marriage supper of the Lamb, the Bridegroom.

Chapter Five

The Mongolian Connection

"**But ye shall receive power, after that the Holy Ghost is come upon you and ye shall be witnesses unto Me both in Jerusalem, and in all Judea, and in Samaria, and unto the uttermost part of the earth.**" Acts 1:8

For many decades Mongolia had been a satellite nation of the Soviet Union, and seemingly the uttermost part of the earth. Then in 1992, with the fall and dismantling of the Soviet Union, Soviet occupation of Mongolia was over and the Russians all left the country. While under the Soviet control nearly all religious worship was banned and churches and temples closed.

But long before the Soviet pullout, the Lord was putting His plan into action, bringing together pieces of a giant puzzle to prepare His bride in Mongolia.

In the early 1980s Herb and Lola, from Montana, took a trip to Israel with a tour group. One of the places they visited was the Galtronics Electronic factory in Tiberias, managed by Ken and Margie Crowell. They were impressed with the work of Ken and Margie and formed a relationship, staying in touch through the years. When the ministry of Galcom International began in 1989 Herb and Lola became partners and supporters of the ministry. As the ministry of Galcom expanded into dozens of countries Herb and Lola became burdened for the people of Mongolia desiring to see them reached with Christian radio. The first two pieces of God's Mongolian puzzle had made a connection.

In 1990 a young American lady went to Mongolia in "creative access" missions to teach English as a third language (most natives spoke the Mongolian language as well as Russian). She taught English using the Bible as a text book. One of her students was a young man named Batjargal Tuvshintsengel (we will call him Bat) who came from a communist family and whose father was a leader very high in the communist party. This family had been trained under the communist ideology and did not believe in God or any form of religion. They were told that religion was opium for the people and that there was no God.

As Bat studied English from the Bible he was confronted with Jesus and began to question his atheistic belief. For three months he wrestled with Jesus and the Gospel message. After many long late night discussions with the American lady he finally gave his heart to the Lord and accepted Jesus as his savior. His father and the rest of his family were angry with him but he knew he had made the right decision. At this time in 1990 there were only six known Christians in Mongolia. The third and fourth pieces of the puzzle had connected.

In September of 1991 Ken Crowell, Harold Kent and Gary Nelson hosted Tentmakers Missionary Conference in Tampa, Florida. One of those who attended the conference was a young man, Battle Manassas Brown, who had already made several trips to Mongolia providing advice and aid to the new government and taking much needed aid supplies to the people in Mongolia. Battle had met Batjargal who was now running a small tourist agency in Ulaanbaatar. By this time Bat was married and was a member of a small but growing Christian church in Ulaanbaatar. Battle became the fifth piece of the puzzle.

In 1995 Gary and Mary Nelson and Harold Kent were impressed to lead a prayer team into the 10-40 window to pray in gateway cities. (The 10-40 window is an area from 10 degrees north latitude to 40 degrees north latitude and from West Africa to East Asia). This area contained over 75% of the population unreached by the Christian Gospel. As they were coordinating their trip with the Christian Information Network in Colorado Springs they learned that no teams had signed up to go to Ulaanbaatar, the capital city of Mongolia. The team decided to go to Mongolia and pray for God to move in that country. A team of seven began preparations for this prayer journey.

Since we knew that Battle Manassas Brown had made many trips to Mongolia, the team asked him for advice and contacts there. He arranged for Exodus Travel, the travel and tourist company managed by Batjargal, to host us. We arrived in Ulaanbaatar on October 24th, 1995. Bat met us at the airport and hosted us for the seven days of our prayer journey there, having arranged prayer sites for each day.

During the week, Bat shared with us his leading of the Lord to one day have his own radio program or perhaps even his own Christian radio station. Harold and Gary told him that we represented a Christian radio ministry called Galcom International and that perhaps we could help. More pieces of the puzzle were coming together.

About two years later Bat was given a time on state radio for a Christian program. He then made application for a license and frequency for his own family radio station. In May of 2001 a license was awarded for Wind FM 104.5 to go on the air. FEBC provided Batjargal with his first 250-watt transmitter. Galcom International sent Bat some fix-tuned radios as a test and they were a great success.

In 2002 Bat learned that two more cities might open to new family stations and he made application for radio licenses in both Darhan city and Erdenet city. Batjargal realized that it would take several miracles of God to get two more transmitters, towers, studios, and all the other necessary equipment to make this happen. With these three radio stations Bat could potentially reach nearly 70% of the population of Mongolia with the life-giving message of Jesus Christ. He called Galcom International explaining his situation.

Herb and Lola learned of Batjargal's great need and agreed to provide a new 2000-watt transmitter for Ulaanbaatar, making the 250-watt transmitter available to one of the other projects. They also provided funding for 1200 Galcom fix-tuned radios to be distributed into growing suburban ger communities around Ulaanbaatar (a ger is the tent of the nomadic mongolian).

Adding to this story, in mid-2003 Galcom International USA got a call from a missionary lady, Yvonne, who was home on furlough from Mongolia, asking if we had any radios that she could distribute into the Gobi Desert. Yvonne has been in Mongolia for many years as director of a crippled children's home. In 2002, she turned the management of the home over to Mongolian Christians and began asking the Lord for a new assignment. She felt that the Lord impressed her to buy two camels (yes camels) and a tent and begin traveling in the Gobi desert where very poor nomadic Mongols live. She notified her church in Clearwater, Florida, that she needed $2400 to purchase two camels and a tent for this outreach.

When the pastor of this church announced this very unusual missionary offering, hands began popping up all over the congregation and he was able to raise the $2400 for this 68 year old missionary lady to begin this new outreach. Her church also raised money to provide her with Galcom Go-Ye radios tuned to Wind FM to distribute wherever radio reception was available.

Again, God had sent His angel ahead to prepare the way. We could never have imagined our little radios being distributed to ger tents in the Gobi Desert, preaching the Gospel to Mongolian Nomads. Isn't our God good? He wants everyone to hear the life-giving message of Jesus.

We are hoping that the radio signal will reach deep into the Gobi desert so that these nomadic people can hear the good news of eternal salvation.

Although the puzzle is not complete, and permits are still pending for new stations, the pieces are connecting and the picture is starting to take shape. God is calling a portion of His bride from the far reaches of Mongolia, this uttermost part of the earth.

Chapter Six

A Harvest in Moldova

"He that gathers in summer is a wise son, but he that sleeps in harvest is a son that causes shame." Prov. 10:5

Moldova is a nation nestled between Romania and the Ukraine. It had previously been a part of Romania. After separating from the Soviet Union and becoming an independent country in 1990, its economy was in shambles. Even after many years as an independent nation the people of Moldova are still desperately poor and struggling for existence. And what is even worse, they have not had the Gospel message preached for decades. Under communism the light of the Gospel had virtually gone out.

But a young man named Florin Pindic'Blaj found the Lord and became burdened for the people of Moldova and Romania. In 1985 he immigrated to the United States and started a Christian organization, Little Samaritan Mission (LSM), to help his countrymen. He began raising money to provide compassion aid and also to start Christian radio stations in the major cities of Moldova, and more recently in Romania.

Galcom International has been helping Florin spread the Gospel to the people of Moldova and Romania by providing Galcom Go-Ye solar powered radios fix-tuned to LSM Christian radio stations. Since 1993, Little Samaritan Mission has been able to install 12 Christian radio stations in the major cities of Moldova and 15 in the major cities of Romania.

Florin Pindic'Blaj was guest speaker at a Galcom International banquet and presented a compelling message regarding the need for Christian radio to reach the multitudes in Moldova and Romania. He has distributed 14,500 Galcom Go-Ye radios (to date) throughout Moldova, that were donated to his organization by the donor partners of Galcom International. Galcom is now providing Go-Ye radios fix-tuned to the LSM Christian radio stations in Romania. LSM radio stations are linked together by satellite to enable sharing of programs, which can also be heard throughout Moldova and Romania by internet broadcasts.

Florin told us that many people do not have jobs, and even have their electricity and water cut off. Large percentages of the people of Moldova are desperately poor, often not able to heat their homes in the winter or even buy food, let alone buy radios or batteries. Galcom solar powered radios fill a great void and are reaching thousands with the Gospel. With Galcom solar powered radios they can still listen to the Gospel broadcast from LSM Christian radio stations even when electricity is shut off.

In Moldova, many government van drivers (public transport) have the Go-Ye radios sitting on the dashboard of the vans, and have the LSM Christian programs playing all the time for all passengers to hear. Florin said that very often the passengers thank the drivers for being able to listen to the radio.

Florin said that he also distributes these radios in hospitals, schools, and orphanages and even distributes Christmas packages to the poor, which include a Galcom radio.

He said that 40% of the calls coming into the radio stations include requests for more Galcom fix-tuned radios. Galcom's solar powered radios have become something unique in Moldova and are filling a spiritual void. The Gospel of Jesus Christ is bringing fresh bread to the hungry souls there.

Florin told us an interesting story about the durability of the Go-Ye radios. A lady in Moldova had one of our Galcom radios and would listen to it all day long. One day her unsaved husband came home and got angry with her and threw the radio against the wall smashing the case to pieces. She gathered up the pieces and the next day took it to the LSM radio station where they glued it back together and it still worked. Another day her husband again came home and found her listening to the Gospel on the Galcom radio and this time threw it out the third story window and it shattered on the concrete parking lot below. Again she gathered up the pieces and took it back to the LSM radio station where they glued it back together and again it worked like new. Now she keeps it out of sight of her husband. We would like to hear the end of the story someday. Join us in prayer for the salvation of this husband.

Florin gets many letters at his radio stations from listeners. The following are samples of letters from these precious people

Vitalie in Chisinau writes: "I was walking down the street very upset one day when I ran into a friend who gave me a little blue radio which he had promised me. I took it home and began to listen to it. I was on the verge of getting a divorce but God opened my eyes and I did not. I began to understand the truth. I went to visit two of my friends who were also on the verge of divorce. It would take much time to explain how God worked in the hearts of my friends, but the important thing is none of them continued with their divorce. We love God more than ever. May God bless those who provide these wonderful little radios."

Ana in Chisinau writes: "I am an orphan so I cannot afford to buy a radio. But my dream came true through Little Samaritan Mission when you gave me a little blue radio. Now I have a nice little radio that I take with me wherever I go. At the training center to be a professional tailor, my colleagues bring music to listen to. One day I brought my little radio and when my professor heard it she told the other girls to shut off their music because everyone will be listening to my radio. Since then I always bring my radio to school. My teacher has asked me to. She listens with great joy. I am never separated from my radio. It is like the shoes on my feet. At night I put it between my pillows so it will not wake anyone. This is how I guard my treasure. This is all I can express through writing but my joy is really much greater. Thank you."

Imaricica from Colibasi, (16 years old) writes: "I would like to tell you that I am a loyal listener to LSM radio. My grandfather is very old and he is almost blind, but he never leaves his place by the little radio that he received from you. Every day he sits there close to it and doesn't miss any programs. I listen to as many programs as I can and treasure them greatly. I thank you very, very much for your radio station and may God reward you for your efforts. I want to tell you that LSM radio is truly a comfort to the suffering. We will both pray that this work continues."

Ion from Criuleni writes: "One day an old gentleman heard a prophecy that said that there would come a time when the wall would speak. As soon as he heard it he quickly forgot about it. One day he received a small 'Go-Ye' radio and went home, hung it on his wall, and began to listen to it. He then remembered the prophecy. His wall was now speaking God's Word. Realizing that his time was near he repented. Because he was a very old man he was not able to go to Church. His radio filled his vase with oil so that at Christ's return he will be ready. He thanks you very much for this radio through which he came to know God at a very old age."

Igor-Soroca writes: "Hello my dear friend Mr. Florin! I don't have sufficient words to thank you for the gift you have given me, the radio. Two days before Christmas I went to the post office and had a box from LSM Radio station. When I opened the box I was absolutely overwhelmed. It had many beautiful things: school supplies, towel, socks, soap, toothbrush and toothpaste; but the greatest treasure in the box was a tiny radio. With my need for insulin we never have enough to purchase other things I need. My wish was to have a radio so that I can listen to your beautiful programs. Now I have my very own Go-Ye radio. How incredible! I thank you so much and wish you God's many blessings."

Letter mailed to LSM station in Chisinau - Zina and Valeriu are invalids. Valeriu is blind and her husband has had heart attacks and liver disease. They said, "Since we became sick we have had to sell everything we had in order to have money to purchase medicine and food. We didn't have a radio but someone told us to go to LSM and they would give us a radio. I didn't believe it at first, but went anyway and received the wonderful little radio. Amazingly, since the first day we had the radio our spirits lifted. It seems as though each program and sermon on LSM radio is especially for us. I never believed a radio could change our lives so much in 4 months. We have now both received Christ and are so full of Joy."

LSM gave him a radio when he came to the LSM library in Chisinau. After we placed the radio in his hands, he fell to his knees and began to kiss it and hold it to his heart. A few months later he again came to the LSM library to tell us that he can finally see with spiritual eyes. He can see the Lord. He told us that now he has renewed hope because one day he will see Jesus and will also have the honor to see the one who made it possible to receive this tiny radio.

An LSM station worker related this story: "Eugenia works in the Chisineu public library and heard of LSM radio station through her son. She came to visit our radio station and told how her alcoholic husband spends all his money to drink. She is able to buy a little food to live on with her salary. LSM gave her a Galcom radio. She asked how we knew she wanted one. We told her God knew. A few weeks later we received a letter from Eugenia and she told us that through the little radio she had received Christ. Christ has given her new hope and now her life is not so difficult and lonely anymore. Even though there is no church in her area she told us that her office in the library is her church. There at her office the Lord speaks to her everyday through the radio. She has the volume on maximum so that children coming into the library can also listen to LSM radio and hear the Gospel message."

LETTERS FROM LSM RADIO-CHISINAU: The following four testimonies are the result of one radio in the hands of a nurse, Rodica, in the hospital in Chisinau: (see photos)

1. Rodica is a nurse in a hospital in Chisinau who came to know Christ through LSM radio. After receiving the Lord, she asked LSM for a Galcom radio to take to work and share it with her patients. She began giving it to her patients individually for several hours day and night so each of them could listen to Christian Radio. As she made her rounds, so did the little radio, among the patients, always bringing the Good News of Jesus Christ. Through this one radio many patients came to know Christ. She continues her ministry leaving the radio a few hours with each patient.

2. Vladimir was making repairs on his house when flammable liquid fell and ignited causing 3rd degree burns over 55% of his body. Rodica would give him the radio at night to listen to the Gospel and it was his only source of comfort. Rodica contacted LSM and they came to visit Vladimir and gave him his own radio. He kept it next to his ear night and day. He was so thankful he began to cry. When LSM returned 4 days later Vladimir had died listening to the Gospel on his little Galcom radio.

3. Vasile had an accident at work when a bucket of hot tar fell and burned his body. Like the other patients, his desire was to have the Galcom radio during the long nights of suffering. When he received the radio he was overjoyed and even said that maybe it was meant for him to end up here in the hospital

because now he was able to receive this precious message of the Gospel.

4. Anatol is an electrician who was electrocuted by accident. He lost both arms as well as his hopes and dreams for the future. He said, "My nurse Rodica told me that only God will help me pass this difficult time of my life. From that moment on I started to listen to LSM radio and I realized that I am just passing through this life and that I have an immortal soul and must take care of it. I still have things to learn but I know that this little radio will help me to understand everything I need to know."

Nurse Rodica keeps sharing her one Galcom radio with the many patients that come into her ward for medical care. Through this one radio the good news is made available to many desperate people.

Many of the letters received at LSM radio stations are from people who are too poor to own a radio and are requesting one from the Christian station in their city. These requests are filled when radios are available. Following are two of those letters:

Olga Tighina writes: "I really love LSM radio. Because I don't have my own radio, I go to my neighbor's home to listen to the broadcast. They don't like it when I use their radio and for a fee they let me listen. My wish is to purchase my own radio but I cannot find one that I can afford. Kindly, I plead with you to please send me a small radio."

Well, Olga got her radio. One donor sent $20 specifically for a radio for Olga and we made sure she got one. Olga had paid her neighbors 10 lei, approximately $1, to listen to the Christian station on their radio. It was extremely difficult for her as 10 lei was a lot of money but she sacrificed to have the opportunity to listen to the Gospel. Now she no longer has to pay. She has a Galcom radio of her own and can listen to LSM broadcasts anytime.

Domnica Orhei writes: "I have decided to write you with a request. I am 68 years old and with God's help I live alone. I am an Orthodox Christian but it doesn't matter because the main thing is that Jesus Christ is in my heart and soul. I have an old radio that used to receive a very weak signal from LSM radio. I enjoyed very much listening to God's word all day and all night, but since January 2001 I have not been able to pick up a signal. I would like so much to continue to listen to your broadcast." (Florin made sure that Domnica got her own radio.)

Quote from Florin Pindic'Blaj in a letter to Galcom International and our supporters:
"Once again, thank you for your collaboration with Little Samaritan Mission in broadcasting the Word of God to people of Moldova and Romania. The portable radios have had such an impact in the lives of the people of Moldova. One little girl told us that after she received her much desired radio, she became so attached to it that at night she would put it under her pillow so it wouldn't wake anyone in the house and she continued to listen to the broadcast until she fell asleep. These portable radios are truly a treasure for these people. We have received another story of a blind man who lives alone and listens to LSM on his portable radio. Since he has no watch and cannot see, he uses the hourly announcements from the broadcast to know when to eat and sleep."

Florin states, "Although it is impossible to recount all the stories of people who have been touched by Galcom radios, I want you to know that these little blue radios have changed people's lives. The generous supporters of Galcom have enabled us to give radios to people who would never be able to afford such "luxury."

Florin Pindic'Blaj is that wise son of Proverbs 10 who is bringing in the harvest in Moldova and Romania in this season of the great harvest.

Chapter Seven

Wings of Mercy - Mexico

"**And I saw another angel (messenger) fly in the midst of heaven, having the everlasting Gospel to preach (broadcast) unto them that dwell on the earth, and to every nation, and kindred, and tongue and people.**" Rev. 14:6 (parenthesis added).

The Mexican people are among the most open to the Gospel message, receiving the Word of God gladly. The Lord has reaped a great harvest among the people groups of Mexico using the Go-Ye radios.

Galcom radios have been distributed by various means, some through the Voice of Hope Radio station, some by mission teams going into Mexico, some by other ministries in Mexico like Victorious Christian Harvesters, and some have even been dropped with parachutes by missionaries Jerry Witt, Jerry Wiley and Alex Fedorenko on missions called Wings of Mercy.

The July 2000 edition of Charisma Magazine had a special feature article entitled "On Wings Of Mercy". It tells how missionaries there distributed Galcom radios to unreached Indians in remote mountains by dropping the radios out of airplanes attached to small parachutes. It is encouraging to see Go-Ye radios reaching these people in this way. Like many other ministries who distribute our Go-Ye radios, Witt, Wiley and Fedorenko are reaching people who could not be reached in any other way.

The following testimonies show the result of Galcom radio distribution in Mexico:

The field of aviation ministry isn't glamorous, although some of Witt's friends call him the Indiana Jones of Mexico. The work is daunting and dangerous. Witt says, "People have a romantic view of aviation, but not many want to rough it and pay the price to work in such difficult areas."

But the challenges don't deter these missionary pilots from dropping their precious cargo, one radio at a time, as they work in the mountains as high as 13,000 feet and in canyons twice as deep as the Grand Canyon. In some remote locations the Indians live in caves and wear loincloths. Witt comments, "When you fly and see these huge canyons littered with thousands of huts and trails going up and down canyon walls, you realize it will take more than conventional transportation to reach these people."

It was the Spanish conquest under Hernando Cortes in 1520 that drove these Indians out of the flatlands into these rugged mountains. Witt and his fellow mission pilots have dropped thousands of radios to the Indians living in these regions.

A minister in Mexico City sent us a report on radio drops in the mountains in the State of Zacatacus. At least 40 Galcom radios were parachuted into a very rugged remote area to a tribe of Huichole (wee-chole) Indians. This tribe of 20,000 live in one of the most remote areas of Mexico.

Among the 40 families that received radios by parachute drop, over 300 family members have come to know Christ as their personal Savior. They have put away their animistic rituals causing havoc among the village leaders with concerns that they are falling away from tribal traditions. But these 300 new believers have taken a firm stand for Christ and are growing in the Lord daily, listening to the Gospel daily on these 40 Galcom Go-Ye solar radios.

Jerry Witt sent us this report after a month of radio airdrops in Mexico: "I am happy to report that we've concluded the first programmed dropping run for this year successfully, and safely. HALLELUJAH! Jerry Wiley was able to come over in his plane and help out on Thursday and Friday, which I'm grateful for, since this represents an extra 225 radios dropped. The total radios dropped this week were 680, which represents an investment of nearly $15,000 dollars. WHOA! God is faithful !!! And we're so grateful to Galcom Ministries, Allan McGuirl (Canada) and Gary Nelson (USA) for their burden to provide the radios for this project.

The reward is in seeing with my own eyes, children, women and men run to get the radio, open the box ... and wave at our airplane as we pass overhead, thanking us for the gift. And we ... well ... we KNOW what will be sounding in those remote little mountain huts in a couple of hours ... PRAISE AND WORSHIP TO OUR KING JESUS!

This thrills us to our innermost being and is our reward for the many hours and dollars invested in the boxes, parachutes, manual labor, long flying hours fighting in those deep canyons, and maintenance on the planes to make it happen in a safe manner. NOW WHAT??? Well .. we have three more dropping runs scheduled for this year. We expect to finish dropping around 2000 radios by the time this year is out. Pray for the safety, finances, and materials needed for this to happen. "
SO THAT ALL MAY HEAR!!!   JERRY D. WITT JR.

Mexico: Missionary Threatened with a gun

The mission work among the Indians of the Central Mountains of Mexico can be dangerous. On one trip into the mountains Jerry Witt was in a Tepehuan Indian village ministering in a church that he had planted. While in the village square a group of drunken men came into the camp and began a brawl. One of the men pulled out a gun and began cocking the gun and pulling the trigger. He cocked the gun on four different occasions, and shot at people (twice point blank). Hundreds of people including children were around, and Jerry expected to hear the gun go off any second and be involved in a desperate rush to evacuate wounded (or possibly dead). But four times, all they heard was a "click". Jerry proceeded to jump the young man, take his gun away, and send the group away without the gun.

These things didn't surprise Jerry since he was accustomed to seeing all manner of attacks against the work of the Lord in the field. What was surprising was the authority and respect that Eleodoro and Margarito (the Tepehuan pastors at the village) as well as the Christians have gained with the government officials there.

As Jerry told the story: "I delivered the gun to the authorities and they excitedly shouted at us, upset at the conduct of the young men in their Christian Village, and reprimanded us for not tying them up and bringing them into their headquarters bound so they could give them a whipping and a fine. Wow! Times have changed! Now the Christians in the region (estimated to be over 600 at this point) have gained a very healthy vote in the community and respect due to the testimonies of God's wonders being worked in their lives."

God is so good and protects His people. Jerry has had his life threatened many times and God has always protected him from harm.

Another story relayed to us by Jerry Witt

"Ten Huichole Indians from Puerto Guamuchil showed up at the Springs of Hope ministry base of Pastor Lorenzo and listened to the Gospel without accepting it. Later, Pastor Lorenzo's evangelists were traveling through the mountains and came across their community and were promptly chased away, but not before the villagers had requested a Galcom radio that one of the team members was carrying.

"Several weeks later the evangelists returned there and the head chief and witch doctor invited them into the community and requested that they share this Gospel that they had heard on the radio to the entire village. Over 30 of them readily accepted Christ as their personal savior and asked to be baptized. They did not want to go to the Springs of Hope for baptism, but wanted to be baptized in the river near their town. They desired their public confession to be known to all of their Huichole neighbors by being witnesses to their water baptism - HALLELUJAH.

"When village leaders come to the Lord it affects the entire village. The radio is a tool that God uses to break down the resistance to the Gospel. Pastor Lorenzo specifically asked that I thank those who have made this tool available to them. So let me add my thanks to all of you who have helped reach this village for Christ."

Huichole Indians of Las Latas

An article in a major nation-wide newspaper discussed Galcom's work among the Huichole Indians of Mexico. Here is a portion of the article:

"For centuries, the Huicholes have forged an almost impenetrable wall between themselves and the outside world, enabling them to preserve their culture and animist religion. If their famous hostility toward outsiders was not enough to ward off would-be invaders, the Huicholes sought refuge in rugged land, a region of scorpion-infested canyons the size of North Carolina.

But that isolation hasn't kept out these missionaries, who are air dropping solar powered radios fix-tuned to Christian Evangelical radio stations. Several hundred Huichole have converted to Christianity since this missionary activity started. Over 3000 solar fix-tuned radios have been distributed into the mountains of Central Mexico, many into Huichole territory. The project is part of a global mission sponsored by Galcom International."

This article is encouraging to us that our efforts are being noticed even outside the Christian community. The article did verify that hundreds of Huicholes are being converted in this rapidly growing body of believers. To God be the Glory. We will keep serving Him!

Radios lead Indians out of the past

Rafael de la Cruz was tending his crops in this remote Huichole Indian community one morning when he heard the strange sound of a propeller engine echoing off the canyon walls. Moments later, a small airplane swooped down over the main ceremonial plaza and began dropping cardboard packages lashed to tiny parachutes. "It was like we were being invaded," the farmer said of the falling bundles, which contained solar powered, fix-tuned radios tuned to Christian radio. De la Cruz and other Huichole leaders now point to the day when things changed for their remote community.

With aircraft and parachutes, Wings of Mercy has been able to penetrate the geographic and cultural walls and introduce the Gospel using the Galcom Go-Ye radios air dropped right to their mountain front doors.

Miracles in the mountains
In addition to air dropping radios, missionaries Witt and Wiley often land at remote mountain air strips and hand deliver boxes of fix-tuned radios as well as take medical missionary teams into these villages. One medical mission team was spending time in prayer on the side of a mountain before beginning the medical work. Suddenly the mountain became holy ground. Wiley whispered a prayer in faith, "Lord, capture us in Your glory like You did Moses on Mount Sinai." Wiley then looked down and saw fog emerge from the backside of the mountain, come straight up toward them, and cover the whole top of the mountain. Everyone was impacted by the presence of the Lord. Wiley said, "There wasn't a dry eye. People fell on their faces and stayed there for about 15 minutes. As fast as the cloud came in it just melted away." Although the glory was only for those moments, God's blessings have continued.

Despite the risks, these missionary pilots have dedicated their lives to evangelizing these indigenous people that the Mexican society has forsaken.

Other projects in Mexico

We had recently sent Go-Ye radios to Pastor Harold Link in San Diego, fix-tuned to a Christian FM radio station in Yuma, Arizona, broadcasting into the Mexicali Valley. Pastor Lind distributed the radios to poor people throughout this valley. The FM radios have a string antenna that is often used to carry the radio around the neck.

A migrant laborer working in a lettuce field had a Go-Ye radio around his neck and was listening to a Gospel program as he picked lettuce. After a while the man working next to him asked him about the program and he told him that it was a program about Jesus. After about another half hour, he inquired again asking how he could come to know this Jesus. The Christian man then led his co-worker to the Lord out in the lettuce field. The Gospel message does not return void, but it does accomplish the purpose for which it is sent.

The missionaries working among the people of this valley have received many letters of testimony about how the Word of God is being spread via radio. Here are excerpts from two of those letters translated from Spanish:

From a man from Oaxaca in Mexicali Valley: "By means of this letter I solicit from you a radio from station FM 91.9, which is a great blessing for my life. This is by means of Bible teaching and praises directed to our Savior and Lord Jesus, who has motivated my life by means of this station. I give thanks to God who put the desire in your heart to give me a radio, one which I can carry wherever I want to go, so that I can learn more about God right away. Thank you for this blessing."

From a wife and mother from the Tamaulipas area of Mexicali: "The radio is very practical. I like it because it has helped me in my married life and in learning how to correct my children. It comes with very practical counsel. It has also helped me to speak to others about the Word of God and to tell about my testimony. I give thanks to God for this ministry and I understand how the Lord moves ahead as the Word of God is extended. It is entering into already hardened hearts and is changing them. I pray the Lord that He continue providing for this ministry."

Other Christian workers who also had the radios around their necks broadcasting the Gospel were stopped by co-workers and asked how they could get a radio so they could listen to the Gospel station. The missionaries working among these people are looking forward to receiving many more Galcom radios to be distributed in Mexico.

The Rio Grande Bible Institute, located in Southern Texas in the town of Edinburg, has distributed 1000 Go-Ye fix-tuned radios to unreached people just across the border in Mexico. The Rio Grande Bible Institute Christian station signal reaches well into Mexico and even to some small off shore islands. In these areas of Mexico only the women and children attend church. The men do not want anything to do with religion, and are indeed an unreached culture. The Lord put on the hearts of the people of Rio Grande Bible Institute to give the solar powered radios to these men as gifts. They will listen to a radio and the Gospel WILL impact them.

On one small off shore island within the broadcast range of the Rio Grande Station there is no electricity or phone service. During a recent hurricane, over a hundred fishermen from this island perished while fishing during the storm. Since they did not have radios they were unaware that the storm was approaching the island. Now the island has hundreds of widows and fatherless children. The Rio Grande Bible Institute donated some of the radios to this island fishing village so they can hear the Gospel by solar powered radio even without electricity. God provides the solar power free. This is an open door to take the Gospel by radio to these unreached people who would not be reached in any other way.

Harold & JoAnn Kent-Founders    Ken & Margie Crowell-Founders

Allan & Florie McGuirl-Founders    Gary & Mary Nelson-Galcom USA

Elder man saved - El Salvador    Gypsie woman listens - Bulgaria

Young lady -Bolivia, loves Go-Ye   Man in Hyderabad, India

Nurse Rodica - Moldova 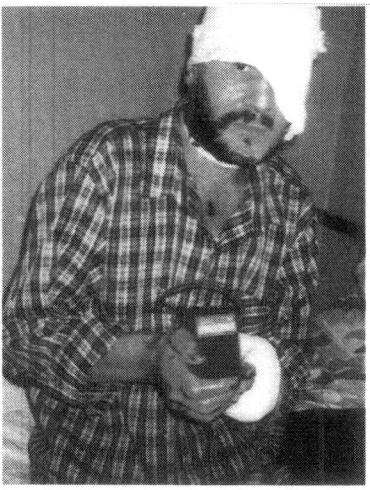   Hospital burn patient - Moldova

Distributing radios - Venezuela    Albanian man happy with radio

Young Inca Mother - Peru    Native Man-Guatemala with radio

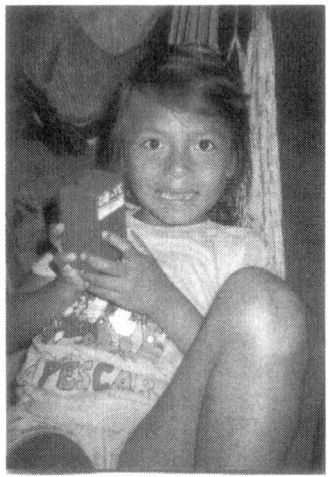

Child in Ecuador hears Gospel    Ixil woman & child-Guatemala

Chapter Eight

White unto Harvest Bolivia- <u>Lugar Blanco</u>

"Say ye not, 'There are yet four months and then cometh the harvest?' Behold I say unto you lift up your eyes and look on the fields for they are white already to harvest." **John 4:35**

Like people in so many other countries of South America, the people of Bolivia need to hear the Gospel. One of the ministries that we have been working with in Bolivia is Living Truth Through Radio Ministries. Bob and Beth White, along with other ministries, have helped establish Mosoj Chaski Christian radio station in Cochabamba, Bolivia. This station broadcasts on short wave in the Quechua language to the over 2 million Quechua Indians in this country. Bob and Beth have moved to Bolivia to live among the Quechua and reap mission fields there that are white unto harvest.

The following is a letter from Beth White expressing the need for the Gospel to be spread by Galcom radios in Bolivia

"The need for the Galcom radios in Bolivia is for reaching the lost sheep of the Quechua people. The Quechua are descended from the Incas. They are an agrarian people, making their living by the sweat of their brow. They are subsistence farmers as they have no viable means of getting the products they produce to market. Their roads are unimproved, since local communities are responsible for keeping up the roads which are high, winding, one lane dirt passages, (really foot trails), hanging precariously on the side of the mountains. The Quechua, with the exception of a few, have no automobiles. A few wealthier Quechua have a donkey or two. The majority of the Quechua we know have no pack animals and no transportation. Their communities are without electricity, and their homes are without running water; therefore, solar powered radios are essential.

"A Christian broadcast radio station, Mosaj Chaski, became a reality about three years ago. Three mission organizations came together to fund and build it. It only broadcasts in the Quechua language and the programming is of Christian content. These broadcasts are essential for reaching the mostly illiterate population.

"These little Go-Ye radios are, in many cases, their only means of hearing the Gospel of Jesus Christ. They cannot afford a radio for their homes which are made of adobe brick and thatch. They cannot read a Bible and for them to be able to hear the Gospel in their own language is a dream come true. I wish I had statistics for you to let you know how many Quechua have come to receive the saving grace of Jesus only through listening to their Go-Ye radio. I do not have those statistics as the Quechua are spread out over thousands of square miles throughout the Andes Mountains which rise to above 18,000 feet.

"Their existence is one of the most base I have ever seen, read, or heard about comparable only to Haiti and Ghana, in the Western Hemisphere. I have been there, worked there, served there and in May I am going to be living there, full time, as a missionary. These little red Go-Ye radios are a true gift from God.

They are helping us in spreading the Gospel of Jesus faster than we could any other way. The people will walk many miles to get one radio. We can't do radio airdrops as small planes do not fly that high.

"Why do we need the radios? We need them to let the Quechua know that even though their own government has forgotten them, even though the rest of the world looks down on them and thinks they are not worth being helped, that there is One who loves them as much as all other humankind. He wants them to receive the same saving grace that is offered to other oppressed and afflicted people throughout the world. There is One who wants them to know that He too walked everywhere He went as they do, but as He walked He carried their sins, and His final miles were a walk to the cross of Calvary so that they could have eternal life with Him. This is the reason the Quechua need the Go Ye radios."

Bob White, of Living Truth Through Radio Ministries, writes

"About a year ago Living Truth ordered 500 Go-Ye radios, with solar panels, tuned to 3310 kHz. We received them and finally were able to take them to Bolivia. We have been well pleased with the performance and especially pleased by how well they are received by the mountain people of the Andes (the Quechua, descendants of the Incan empire).

"They are the perfect radio for this application for two reasons: One is that they are fix-tuned only to the Mosoj Chaski Radio Station and therefore have no 'street' value.

"Secondly, the rechargeable batteries. The Quechua are very poor and even the few who have shortwave radios use them very sparingly due to the difficulty and cost of obtaining replacement batteries. The Go-Ye radios are therefore very popular in these extremely rural areas. Thank you for this service Galcom is providing in the Lord's harvest field here in Bolivia. These radios are reaching people who would not be reached without them.

"In addition, the Quechua can continue to learn and grow spiritually in all of their villages wherever they live. We missionaries cannot continuously be in all of those difficult places, but the radio waves and God's Holy Spirit can be.

"One of the Christian programs that has become very popular is "Bible Story Time". During this program aired twice a day, the station presents the Bible in story form and in a chronological sequence. Over two million of the Quechuas are illiterate, therefore story telling has been a way of life and their story telling abilities have become very well honed. They can repeat stories nearly word for word in great detail.

"They love to hear the Bible stories and all gather around the Galcom radios during Bible Story Time and listen with great attention. Often a dozen or more will be seen crowded around one Go-Ye radio listening to the Bible stories. At the end of each program they are told to go out and tell the story to other people. They are all very eager to retell these stories over and over again. One lady who works in a craft shop tells the Bible stories to her co-workers and many have come to the Lord as a result. This is happening in many Quechua villages.

"In the past the country of Bolivia has not offered schooling in the Quechua language, even though there are 2.2 million Indians who only speak the Quechua language. Radio Mosoj Chaski has begun literacy classes over the radio in the Quechua language. This is opening up a new door for education to these people which they have long desired, to be able to learn to read and write.

"Now villages throughout Bolivia are asking the radio station for Galcom radios so they can receive the literacy training. One village leader asked for 40 Go-Ye radios for the people of his village because of this literacy training program. As they listen to this Christian station for the literacy training they also hear the Good News of Jesus Christ. They are motivated to listen and now, in addition to learning to read, they are also learning about salvation through Jesus Christ."

Dionicio and Arminda are native pastors among the Quechua Indians of Bolivia and share the following testimonies
"The little solar powered, battery operated Go-Ye radios are making such a great impact on the lives of the Quechua Indians.

"An old Quechua man showed up at our Christian center one spring day. He was out of breath, hungry and tired. It seems he had been walking all morning in an attempt to reach the center so that he could get a "Go-Ye" radio and return to his village before nightfall. He told Dionicio that his wife was very angry with him. Somehow he had managed to lose their radio. His wife had ordered him out of the house and told him not to return home without one! He sounded desperate and as he poured out his need before Dionicio he opened his blanket which he carried on his back, tied around his narrow shoulders, and displayed an array of different types of potatoes and some eggs that he was offering in exchange for a new radio."

"Dionico stated this story with us as a way of reminding us of the need to continue bringing the Go-Ye radios into Bolivia for distribution to the Quechua people. This man and his wife are already believers but there are well over 1.8 million Quechua who have not accepted Christ as their Savior. These little radios have the benefit of taking the message of Jesus to people who would never cross the threshold of a church but who will listen to the Christian music and the Bible lessons taught in their native language.

"Another day when we walked into the office at the Center we found a man waiting in a chair all alone. The man had walked eight hours to come to the Center. It seems he had heard, on one of the Go-Ye radios, that at our Center he could learn more about Jesus. He had walked all that way to not only ask questions about Jesus but also to ask Dionicio and Arminda to come to his village (which had no believers) to tell the villagers about Jesus. A village in Bolivia without a church is called a "lugar blanco" (white area) and comes from the scripture in John 4:35 "The fields being white unto harvest."

Another report from Bob and Beth White
"When we first visited Valentine and his wife Dominga, Valentine was seated in the doorway of his adobe house with his trousers pulled down around his thighs. He had been seriously injured in a fight while drinking three months earlier. He had been in a coma but came out of it and now needed crutches to walk. Hanging around his neck was an amulet that had been given to him by the local yatiri (witch doctor) as a cure for his injured leg. Only a week earlier Valentine and Dominga had accepted Jesus as

Lord and Savior. We explained that they no longer needed the amulets from the yatiri, but they could rely on God for all their needs.

"On our second visit they had been drinking, but we shared a Bible lesson with Valentine and Dominga and gave them a solar powered Galcom radio tuned to Mosoj Chaski radio station. The next visit they were so happy to see us. They had been listening to the radio and were very encouraged by what they were hearing. Valentine told us that the programming began every morning at 4:00 AM with prayer and he had begun praying. He told us that they loved the music and sang along with the Christian songs. They also listen to the Bible stories and were able to answer questions about them.

"We cannot be with all the believers to encourage them in their faith, but with the gift of a "Go-Ye" radio people like Valentine and Dominga can hear the Word of God every day. Having this little radio helps them to stop drinking and resist the advances of the yatiri. It gives them a deeper understanding of their new faith and they learn more clearly what it means to be a disciple of Jesus. Thank you for your ministry. The radios are being used well and the people love having them.

"We hope you can help us establish a low powered radio station for the town of Llallagua, Bolivia. This is a mining town in Bolivia, high in the Andes at an elevation of close to 14,000 feet. It is cold there most of the year. Four denominations wish to partner together to spread the Gospel through that area. The pastors of the churches will do the programming. The miners are some of the most lost people in Bolivia. With their pay which we've been told amounts to about $17 a week, they purchase alcohol and seek out prostitutes. We have seen how they live and it is a very sad existence. The churches have already obtained an FM frequency." (Another open door for God's Word through radio).

Through the ministry of Bob and Beth White, Galcom International and others who are working in the white harvest fields of Bolivia, the Quechua are coming to the Lord and being set free from generations of bondage. The harvest there is great.

Chapter Nine
Building in Latin America

"Yea, so have I strived to preach the Gospel, not where Christ was named, lest I should build upon another man's foundation." Rom. 15:20

Guatemala

Allan McGuirl, Galcom's International Director in Canada, attended the annual 2003 COICOM Conference held in Guatemala City. While there he stayed with Rev. Fausto Cebeira and his wife Meriam who operate two Christian radio stations in Guatemala. Galcom had donated Go-Ye fix-tuned radios to this ministry and the Cebeiras shared several testimonies with Allan.

They told of a man who had walked four hours to the radio station to share that he had received a Galcom radio and by listening to the Gospel accepted Christ as his Savior. He was never able to get to town before, and he wanted to personally thank the people who were responsible for what the Lord had done in his life. He said that he listens to the radio then goes to other villages and preaches to them, sometimes walking 12 hours to preach the Word in distant surrounding villages.

He is grateful that the radios are solar powered and don't need battery replacement. After using his radio for four years it still works perfectly. This man preaches among the IXIL Indian people and estimates that there are now 150,000 IXIL believers spread out over this remote region. The radio ministry is reaching this tribe for Christ.

In Rio Azol, Guatemala, one radio listener was a well known Mayan witch doctor named Miguel. After listening to his little Galcom radio he came to know Christ as his Savior. He confessed his salvation in church, explaining that it was because of his radio that he had accepted Christ. His life has been drastically changed. Now he is reaching others for Christ.

Julian Escovar is the pastor of a church in the small town of Biscanal, Guatemala. His town is four hours by foot from the IXIL radio station in Guatemala. Pastor Julian recently contacted us to say that "there are many small towns and counties that are difficult to maintain contact with; the people have no roads and there is no electricity. We have to walk two days to reach these people. These small radios are the only link for them to hear the message of salvation. Thank you for the gift of the radios."

Another church in Guatemala has asked us to help them establish a low powered Christian radio station there, to reach the Pocomchi Indians. As Romans 15:20 states, we are endeavoring to take the Gospel to where Christ is not known.

We recently sent Go-Ye radios to a new radio station in Jocotan, Guatemala. Here is a report we just received back from them
"We are delighted to be able to tell you that Radio Alegria went on the air early in February. This is in the Chorti Indian area that we began visiting about 14 months ago. There has been an overwhelming response, especially from the Indians who can now listen to the Gospel in their own

language. Many have already traveled to Jocotan to visit the radio station and even bring a small offering. We were able to give some of your solar powered radios tuned to 97.5 FM to village leaders in 10 villages where we are working. They were delighted. Thank our precious Lord and thank you and the people of Galcom for this extraordinary advance among this language group. Would you please send us 1000 more solar radios for these poorest of the poor people." Gratefully, Fausto & Mariam Cebeira

Honduras

A Canadian Pastor was visiting a mission church in Honduras and had taken along some Galcom Go-Ye radios tuned to the local Christian station in the area. One day he was in town with the local pastor and saw an elderly man sitting on a park bench. The Canadian pastor said "Let's go and tell this elderly man about the Lord." The pastor replied, "Don't waste your time. I have been trying to share the Lord with him for years but he is stubborn. He says I don't need your Jesus."

The Canadian pastor went over to the elderly man and asked him if he would accept a gift from the people of Canada. He gladly accepted a free Galcom Go-Ye radio fix-tuned to the local Christian radio station. A few days later the two pastors were again in town and saw the elderly man holding the radio to his ear intently listening to the program with a glow on his face. They went over and asked him how he was doing. The man explained that he had accepted the Lord while listening to the Gospel on the radio and was already sharing the good news with some of his friends.

Radio is a wonderful tool that can cross racial, cultural or religious barriers with the good news of salvation through Jesus Christ.

Chapter Ten

A Harvest in South America

When Jesus sent 70 of His disciples out to minister He said to them, "The harvest truly is great, but the laborers are few, pray ye therefore the Lord of the harvest that He would send forth laborers into His harvest." Luke 10:2

In addition to the radio ministry in Bolivia God is raising up and sending out many harvest laborers throughout countries in South America as a result of Christian Radio. Although in many areas this work is just beginning, it is bringing in a great harvest.

La Morita, Venezuela

Zabdiel Arenas, of El Renuevo Radio ministries, where Galcom had installed a radio station, shared many great testimonies as a result of this radio ministry in Venezuela. He told of a 17 year old lady who came to know the Lord by listening to the Gospel on a Go-Ye radio and through her radiant testimony her whole family plus two other neighbor families also came to know Christ. Zabdiel arranged for her to attend a Bible Institute and she will be the first missionary sent to work among her people.

Zabdiel also told us of a local Catholic Priest who announced to his people, "I am using material from the programs over the radio in my sermons. I encourage all of you to listen to this Christian radio station."

When the Christian radio station was installed by Galcom, the team left 1000 Go-Ye fix-tuned radios with the station personnel to distribute. They went up the river distributing Galcom radios to people in villages as they went.

These villagers receive supplies by a barge that has an infrequent and irregular schedule. People listen to the Christian station intently to be informed of the supply barge schedule announced over the radio. But all the while they hear Christian music, Bible readings and Gospel messages. This Christian radio station plays Bible cassettes, Christian music and preaching in their language over 90% of the time. Many are hearing the Gospel for the first time. They are challenged with the message of Jesus Christ and must make a decision about Him if they desire to continue to listen to the radio for news of the supply barge. Pray that they will receive the good news of salvation with joy.

Maripa

Allan McGuirl and Bruce Foreman from the Galcom office in Canada took a group of ladies from Peoples Church, Toronto, to install a Christian radio station in Maripa, Venezuela. During their stay they primed, painted and erected an 80 foot tower, started the foundation for the studio and distributed Go-Ye radios that they had taken with them. The transmitter and studio were set up in a temporary structure until the studio building was completed.

There was great excitement in the whole area when the transmitter was turned on and the station began broadcasting Christian music and Gospel messages. It was thrilling to see several of the local people come to know Christ immediately after the station started broadcasting.

Palmarito, Venezuela
　　In 2001 Allan McGuirl took a team of 20 to Palmarito, Venezuela, where he and the team installed a Christian radio station. This station, in a community of 30,000 people, was the first of three new radio projects in Venezuela that had been approved. The other two projects are Puerto Ayacucho, 300,000 people and Achaguaqs, 45,000 people. These are all in areas where there is no Christian radio and very little Christian witness. We feel these and more Christian stations will have a dramatic impact on Venezuela.

　　The team of 20 not only delivered Go-Ye radios but also set up the transmitter and transmission tower and constructed the studio building. Quite a project for a nine day trip.

The Amazon region of Brazil
　　Sixty three people in Brazil dedicated a month to preaching the Gospel to children who live along the borders of the Amazon River during what was called Odyssey on the Amazon. The group traveled in ten boats, distributing 3000 Galcom radios preset to Trans World Radio and HCJB. They reached 8422 children and 3576 received Jesus Christ as Savior.

Paraguay
Pastor Jose Holowaty of Iglesia Biblica Missionera operates a Christian radio station in his church in Paraguay. A couple of years ago Pastor Holowaty sent several hundred Galcom Go-Ye radios into the Western Interior of Paraguay. After two years he sent a church planting team to that region to start a church. At the first meeting over 250 people showed up for the service and they were all believers. The team asked them how they had come to know the

Lord as no pastors, evangelists, or missionaries had yet been to the area. The people said they had come to know the Lord listening to the Go-Ye radios and had been waiting for someone to come and start a church for them.

Pastor Holowaty writes, "There are many testimonies which we could include in this report, of people who have come out of the darkness of sin and have come to know the Lord Jesus Christ as personal Savior. I encourage the brethren who collaborate in the fabrication of these little devices to continue to do so, 'knowing that their work in The Lord is not in vain.' (1 Cor. 15:58)

"Many thanks for the great help in providing these little radio receivers. Owing to the extreme poverty faced by many families desirous of attending our church, we actually pay bus fare to and from certain locations. We presently lease three sized buses so that many of the poor can attend church. The same takes place with the radio receivers. Our station is on the air around the clock. Now, with the help of Galcom, we can provide these poor people with radios as soon as they arrive at our church."

Suriname project - **Background**

About 150 years ago the Dutch sailors were involved in slave trading between Africa and the Americas. Their slave ships would dock in Dutch Guyana, (now Suriname) South America, to sell some of the slaves there. Many of these African slaves escaped and fled deep into the jungle, along a river. There are now about 20,000 descendants of these African slaves living in small scattered villages in the jungle about 100 miles from Georgetown, Suriname.

Our Galcom radio engineer, Dave Casement, working with a Suriname mission and Towers for Jesus, installed a Christian radio station in this jungle location to broadcast the Gospel to these 20,000 people who have never heard about Jesus. Galcom supplied 1000 Go-Ye fix-tuned radios tuned to that station to distribute to this unreached people group. This is a new tribe reached for Christ.

Here is the first report received back from station directors Charles and Brittany Shirley:

"Hey Guys, we have been transmitting for a little over a week now. We have the Gospel being transmitted out to places that defy physics. We are hearing reports of people receiving the station up to 100 miles away in the opposite direction from where the tower is pointed. The reports keep coming in from all over the place. Pretty exciting huh? "

Colombia

Russell Stendal in Colombia is reaching a variety of cultures that are nearly impossible to reach in any way other than Christian radio. Guerrillas, drug growers, paramilitary, Colombian Army, Native Indian tribes and other groups that can receive the Gospel being broadcast on his station. When he placed his first order for Go-Ye radios recently he sent us the following letter:

"Thank you so much for your interest and help. Our radio station is located in Eastern Colombia right in the middle of the war zone. Marxist guerrillas are to the south of our station, right wing illegal paramilitary forces are to the north, the Colombian Army is to the west and the drug growers to the east. To the south there are not any Christian churches. All have been closed by the guerrillas. Our broadcast on short wave covers the entire country of Colombia and extends into Latin America and even Europe at night. There are also fifty or more Indian tribes who listen to the station. Many Christians in these areas have no other means of receiving teaching except our radio station."

God has placed Russell and his radio station in a strategic location to reach these varying cultures. After receiving his first order of Galcom Go-Ye radios he immediately sent us the following report:

"We have already distributed most of the radios you sent us. Many are in the hands of guerrillas. The Lord is using these radios to touch many lives. Eleven guerrillas surrendered to us after listening to our radio station. We were able to send radios to many of their friends and now another group of thirty are ready to surrender. We are also sending radios to many small churches in isolated areas." Russ Stendal

The Lord is sending laborers into the harvest fields of South America has allowed Galcom International to be a part of that labor force, extending the arms of these missionaries around thousands of unreached people.

Chapter Eleven

Reaching into The Caribbean

"**The Lord reigneth; let the earth rejoice; let the multitudes of the isles be glad thereof.**"
Ps. 97:1

Haiti: - Operation Saturation

In the year 2000 we began this project working in conjunction with OMS International and Men For Missions with a goal to saturate Haiti with the Gospel message by Christian radio. This island country, so steeped in witchcraft and voodoo, needs a spiritual breakthrough.

The goal includes the distribution of 200,000 Galcom Go-Ye radios fix-tuned to Christian radio stations operating in Haiti. In the first two years of Operation Saturation we have produced, delivered and distributed 15,000 Galcom radios in Haiti. With these 15,000 radios we are now reaching approximately 315,000 Haitians with the Gospel, 18 hours a day, seven days a week.

The term "saturation" refers to intercession as well as to the radio signal and fix-tuned radios. A new breed of valiant prayer warriors, intercessors, are there fighting the war on their knees. The victory must first be won in the heavenlies.

Results: One of the first towns targeted was Gaudin. Go-Ye radios were distributed to 80% of the households in this village. Here are some testimonies from Gaudin.

Pierre Milien, over 100 years old, had lost his voice and strength, but accepted the Lord listening to a radio given to his household. Praise the Lord he is now ready to meet his Creator and Savior.

Ruben Louis, a 20 year old bricklayer, was closed to the Gospel. One evening he was visiting some friends who were listening to a Galcom radio playing the program "Let The Rocks Cry Out". Ruben's heart was convicted by the Spirit's voice and he immediately knelt down and gave his life to the Lord.

Altagrace, a 76 year old lady, returned home from a long hospital stay unable to walk. She borrowed her neighbor's Go-Ye radio and heard a message that shook her to the core. She called for Christian leaders to come and pray for her. As they prayed she felt herself bathed in joy, was completely transformed and saved. Eight days later she began to walk for the first time in months.

Dozens of testimonies like these are circulating throughout this community. Now nearly the entire town has turned to Christ, including the local witch doctors.

We recently received the following testimony from Haiti where some witch doctors and their families are being reached with Galcom Go-Ye radios. These witch doctors would put curses on missionaries or anyone who tried to tell them about Jesus but they would listen to a radio. After listening to the Christian broadcasting many of these witch doctors gave their lives to the Lord renouncing their past ways.

A Pastor from Haiti who was visiting the States phoned Galcom with a report saying he was delighted about the results of the fix-tuned radios and how they have been used to touch so many lives.

A witch doctor had received a Go-Ye radio and came to know Christ soon afterwards. The radio station heard about it, interviewed the man, put his testimony on the radio and soon after that they heard of two other witch doctors who had come to know Christ as a result.

A blind man was with one of the converted witch doctors who gave him a radio. Shortly after that he came to know Christ and gave the radio to his son and the son also came to salvation in Christ.

This pastor also told us that the radios were being used at Pigion Hospital in Haiti. The nurses were loaning our fix-tuned radios to people who were very ill or who were in the waiting rooms. This pastor tells me that many people have been healed and many have been saved through the radios. We praise God for using us to reach the bride of Christ in Haiti.

Cuba
George Otis shared this amazing testimony while he was with us for our Galcom board meeting. A young man in Cuba who was on Castro's staff had a one week vacation. He lived 25 miles outside Havana and there was nothing for him to do for entertainment during this week. Someone gave him a Galcom fix-tuned radio and he loved the Christian music. But to listen to the music he also had to listen to the Bible readings, the calls to salvation, healing prayers, etc. After listening to this station for a week he was gloriously saved.

His mother had been in a mental hospital for 17 years. He thought that perhaps if he could take this radio to her so she could hear this good news, that it would help her. He took the radio to the mental hospital, turned it on and placed it under her bed. After two and a half weeks, she suddenly sat up in bed and said, "Where am I." She was totally restored. Two days later she was released from the hospital and went home totally healed.

In another case a woman was dying of cancer. Finally, the husband could no longer care for her at home and made a pallet for her on his bicycle to take her to the hospital. It was a 3 hour trip by bicycle. She got hold of a Galcom radio, tuned to the Voice of Hope and was listening to it on the trip. After they had traveled for an hour and a half the message on the radio included prayer for healing. As her husband peddled the bicycle she prayed for healing along with the radio program. She told her husband to stop, turn around and go home as she was totally healed.

A pastor of a large church in Ciego de Avila, Cuba, was in Tampa and visited the Galcom office. He was very excited to learn about the radios going to Cuba and said that people all over Cuba listen to the Christian station. He also took a Galcom Solar PA with him to Cuba, saying he would use it in Church on Sundays since electricity is unreliable. Many Cuban pastors that come to the USA are invited to pre-record Gospel messages for the Cuban people to be aired over the radio. The people in Cuba especially enjoy hearing messages by their own pastors being broadcast over the airwaves in Cuba.

Puerto Rico (Janet Lattrell's prison ministry)

Janet writes: "Thank you so much for your part getting radios to prisoners here in Puerto Rico. The radios are given to those who request them or who are referred to us by loved ones. They are also given or sent to persons that cannot afford to buy them. We get many reports of how people are turning to the Lord, thanks to hearing the Gospel message on the little Go-Ye radios. The staff and supporters of Galcom truly have a great joy and reward in heaven from the many who have met Jesus Christ as Savior through the little radios."

Chapter Twelve

God Rains on The Rain Forest

"For as the rain cometh down, and snow from heaven, and returneth not thither, but watereth the earth and maketh it bring forth and bud, that it may give seed to the sower and bread to the eater; So shall My Word be that goeth forth out of My mouth: it shall not return unto Me void, but it shall accomplish that which I please, and shall prosper in the things whereto I send it." Isa. 55:10 & 11

What is on God's heart for Central Africa?
This chapter's title and scripture verse rang so true because of a series of recent contacts. The Lord began impressing on us in 2003, to add the Pygmies of the rain forest to our long list of unreached people groups that are on His heart. A series of contacts and phone calls made us realize that now must be His time to take the Gospel to these people.

The Pygmies, called the Baka people, are tribal people who live in the dense jungles of The Republic of Congo, Gabon, Cameroon, The Central Republic of Aftica and The Democratic Republic of Congo (formerly Zaire). They live in small villages and subsist from whatever food the animals, insects and vegetation in their area can provide. When the jungle area they are living in can no longer support the village they move to another location. This makes them very hard to find and evangelize. Even when a village is reached, the next time the missionaries go back the village has often moved to another location and cannot be found.

In mid-2003, Galcom began receiving inquiries regarding the people of Central Africa and the possibility of reaching them with Christian radio.

The first such contact was from a missionary who had been working among the Pygmies in The Republic of Congo. He called the Galcom office in Tampa, asking about using Christian radio to reach into the jungles of the Congo Nations where the Pygmies live. This missionary had returned because of family health issues but wanted to go back to Africa and continue to work among the Pygmies. He knew that he could be much more effective with Christian radio.

The second contact was from Global Outreach, which has a mission base in The Republic of Congo. They have asked us to help them install a Christian radio station on their base. A tropical band radio transmitter from their base could reach deep into the rain forest where the Pygmies live.

A third contact has been with Campus Crusade for Christ in Canada. They have a mission base in Cameroon on the Northern Border of The Republic of Congo and wanted to work with Galcom to install an FM and tropical band radio station in that location. This station would reach the large Pygmy village near the mission base with a good clear FM signal but also reach deep into the rain forest with short wave where the nomadic Pygmies roam.

As a result of the above series of calls we realized that it was time for the African Pygmies to hear the Gospel, and the Lord wanted us to be a part of reaping in that harvest field. We said to the Lord; "Okay Lord, we are willing to do it, enable us."

Because of this obvious prompting from the Lord we have develped a project to install a Christian short wave station in Cameroon to reach the Baka people living in the jungles of Southern Cameroon, Southern Central African Republic, East Gabon, Northern Republic of Congo and Western DR Congo. Known as the "forgotten people of the rain forest" they have been neglected in previous mission efforts. There are presently few people attempting to work among the Baka. No one has been able to penetrate the rain forest and follow their nomadic pattern to reach them. Only recently God has stirred the interest of some mission groups to attempt to take the Gospel to the Pygmies.

Once the radio station is set up it will broadcast Biblical teaching, songs, stories from the Bible, plus educational, health and community development training, all in the Baka language. Under the leadership of Pastor Donald Ndichafah, a Cameroonian missionary couple will direct the radio ministry. They will work along side a Baka speaking Pygmy couple who will translate the programs into the Baka language.

We hope to reach entire family units and villages with the Gospel. We will mass distribute the Go-Ye solar powered radios using teams of pygmies that will traverse the rain forest following petty trader routes. Once the family units or villages are reached they will be able to listen to the Gospel programming daily regardless of their nomadic lifestyle.

This is God's time for the Baka Pygmies of Africa. We feel honored to be called by God to be a part of this exciting project to send the rain of God's Word to the people of the rain forest.

Ongoing work in Eastern DR Congo
Richard and Kathy McDonald

In 1994 we had sent to Zaire, now The Democratic Republic of Congo, a radio transmitter in a suitcase and an initial order of three thousand Go-Ye fix-tuned radios. At that time Richard and Kathy McDonald started FM 91.1 "Christian Radio Kahuzi" in Bukavu, The Democratic Republic of Congo, broadcasting the Gospel message to the 10 million people living within the 30 mile listening radius of the station.

Fifty percent of these 10 million people were refugees from Burundi and Rwanda who had fled the genocide occurring in their homelands. Richard and Kathy have expanded their radio outreach in Bukavu, DR Congo, by also putting in a short wave transmitter with the signal reaching hundreds of miles. This short wave signal now reaches into the forest where the Pygmies of East DR Congo live.

The next time we heard from the McDonalds we were amazed to find out that hundreds of thousands of people had been reached with the Gospel and were listening to Christian radio daily with these initial 3000 radios. Only The Lord knows the full impact. God's Word is going forth and is not returning void but is going into fertile soil and accomplishing His purpose for which He sent it.

We have continued to supply Galcom Go-Ye radios to Radio Kahuzi as there are still millions in the listening area that do not have access to a radio. These people still live in refugee camps, many without the basic necessities of life; however, with a solar powered Galcom radio they can get the bread of life by joining a radio club.

Due to the large demand for the fix-tuned radios, Richard and Kathy established a requirement for anyone receiving a radio to have formed a radio club with at least 50 people in the club. In the past 10 years hundreds of these radio clubs have been formed, the largest club having over 800 people in it. The people in these radio clubs share the Galcom radios with many people listening to the Gospel on a single radio at the same time. Using the radio club concept of at least 50 people per club, these radios will reach tens of thousands of unreached people with the Gospel.

Club members correspond with the station regularly and center their personal lives around Christian service in their villages. They find an outreach for witness to friends and family by sending letters to the station that are read during the correspondence program. This program also provides an opportunity to request Christian songs to be directed to friends and family as a way of sharing their faith.

God has blessed the ministry and outreach of each radio club in different ways. Several clubs joined together to reach and win the hearts of street kids, widows and orphans. One of these clubs has become so successful that the new government invited some of the members to the capital city Kinshasa to work with street kids there.

God blessed another radio club which began to work with Pygmies in the jungle. They are teaching the Pygmies to read and write while ministering the Gospel to them. Now the Pygmies have their own radio club and the president of this club, one of the first educated Pygmies, announced his willingness to teach his own people. He is teaching them by having them write letters to Radio Kahuzi and signing their own name. Since listening to Christian radio, some of the Pygmies see the importance of Christian marriage and want to have a Christian wedding. This tribe has never had a Christian marriage.

Other clubs do agriculture, fish farming, animal raising, bee-keeping, tree planting, and road repair. Additionally, many clubs are involved in sanitation projects such as cleaning up their villages and cleaning out ditches. These efforts have reduced the effect of a recent cholera epidemic around Bukavu. Some clubs take up offerings to help its members in distress. There is one girl's club that is training the members to become better future homemakers and Christian parents.

God has blessed the distribution of thousands of solar, fix-tuned radios. This has become a full-time work and ministry itself. Priority has been given to refugees, displaced persons, hospitals and prisons. One man was in prison for beating his wife and pouring hot oil on her in a fit of anger. After some time contemplating what he had done he began to listen to a little Galcom radio he had been given, tuned to Radio Kahuzi, 91.1. That day, while listening to Dr. James Dobson's "Focus on The Family" in French, he was convicted in his heart, confessed his need for a Savior and committed his life to Christ. He was reconciled to his wife and family and is now out of prison and has become an evangelist.

Gila Garaway, wife of the late Noah Garaway, who first designed and produced the Go-Ye radio in

Israel, was recently in Bukavu and told us she was amazed to see Galcom radios everywhere. The man searching her bags at the airport was wearing a Go-Ye radio around his neck listening to Radio Kahuzi. Officers in government offices were listening to Galcom radios as well as the receptionist in the courthouse. Shopkeepers in the big markets had radios on. A woman selling tomatoes was faithfully playing her little Go-Ye radio for all to hear.

One day Richard MacDonald was visiting a camp of UN Peace Keeping Forces of Chinese soldiers. He asked them "What language do you speak, Mandarin?" They replied, "English." Richard was able to give some of the Chinese soldiers Galcom radios to listen to as they worked. We could never have imagined reaching into the Chinese Army with the Gospel in such a unique way, but with God all things are possible.

Is the radio worth the investment of time, money, and personnel? I think you will agree that in the "Third World" radio is sometimes the only means of communicating the Gospel.

From this report from Richard and Kathy McDonald we can see that it is necessary to continue to work in the harvest fields while it is yet day. God is raining His Word on Central Africa and His Word is not returning void.

Chapter Thirteen

All Over Africa God's Spirit is Moving

"After this I beheld, and, lo a great multitude which no man could number of all nations, and kindreds, and peoples and tongues stood before the throne and before the Lamb, clothed in white robes, and palm branches in their hands and cried out with a loud voice saying 'Salvation to our God which sits upon the Throne and unto the Lamb.'" Rev. 7:9-10

Africa has long been known as the Dark Continent. One of the reasons it has been so dark is not because of the color of the people's skin, but because of the spiritual darkness. Now the light of the Gospel is being shown throughout this continent. There are tribes and nations coming to the Lord in Africa. Galcom radios are proclaiming the Gospel to remote places to unreached people.

Burkino Faso

Jim Sawatsky, President of GNI Media Association of Central Africa, sent us this urgent request: "Remember Pastor Henock Hema? If you have been following our news through the years you would know that he was diagnosed with sickle cell anemia and is nearly blind, yet he continues to serve the Lord. Yesterday he came to this city to tell us that God had spoken to him the night before.

"For six hours during the night The Lord came to him in a vision. God wanted him to use radio to reach the older people in their mother tongues. Older people there do not travel or read. There are 14 language groups in his area and many do not have pastors yet. He plans to prepare radio messages in all these languages and somehow get radios to these older folks who have yet to hear the Gospel in their own language." Galcom has supplied Jim Swatsky with Go-Ye radios to meet this need.

Another interesting story occurred recently when one of our supporters, Henning Moe, called with this testimony. He had a lady visitor from Ouagadougou, the capital city of Burkino Faso. She found out Galcom had set up a Christian radio station in her city and was operated by a local church. Henning told this lady that he would order 20 Galcom Go-Ye fix-tuned radios tuned to the Christian radio station in Ouagadougou and send them to her father for her to pick up when she arrived back home. When her father got the radios he didn't save them for his daughter but instead he inadvertently distributed them to unbelievers who could not be reached in any other way.

Apparently her father had three wives, one Christian (who was this lady's mother) and two muslim wives which were his favored wives. So he gave a radio to each of his Muslim wives. He is also a heavy drinker, so he traded the remaining 18 radios to whiskey peddlers for drinks. Now his two muslim wives and 18 whiskey peddlers in Ouagadougou, Burkino Faso are listening to the Gospel on Galcom radios. These people would never go into a Christian church or listen to a missionary or evangelist even if there were one there to talk to them. How could they have been reached in any other way? Henning said, "I think I'll send him another twenty radios."

Testimonies sent by Andrew and Esther Schaeffer in Burkinabe, a Muslim area

"Every evening, around 6:00 p.m., we hear him bicycling down the street on his way to work. Always smiling and ready to greet us, Mr. Lamien, our night guard, enters our compound with his Galcom solar radio around his neck listening to Christian music. While he waters the flowers, cuts the grass, eats his supper and guards our house, he listens to "Focus on the Family" or a message preached by one of our local pastors. He often says to us, "We listen to the radio as much as we can." It seems to us that he listens to it all day long.

Colima Church is located in a neighborhood very resistant to the Gospel. After five years the church has seen very little conversion growth, but in recent days that has changed. Last week, 18 women who were attending a weekly Bible study were each given a Galcom radio. You would have thought they had been given a thousand dollars! A spontaneous cheer was followed by enthusiastic singing. They were so excited to have these radios. As they listen they are discipled through the radio and are allowing neighbors and friends to also hear the Good News and be drawn to Christ. The pastor's wife said, "We aren't quite sure what to do with all these women that have accepted Christ."

He has a small lean-to outside of a courtyard and all day long he repairs flip-flops and broken sandals. He also makes brooms out of dry weeds that he collects outside the city. Hanging on a pole in his shop is a small solar radio, playing Christian music and proclaiming God's Word. His stand is located right next to a busy market area, so there are always people walking by. Some stop and listen to the radio, pulling up a stool next to the radio. Others pause, listen, then walk on. The people who visit this market are hearing the Gospel in this natural and non offensive way.

The director of the radio station in Burkinabe said, "These radios have changed everything, especially for young people. They can listen on their breaks from class. Christian radio has revolutionized our city." Andrew and Esther add this comment, "We are so thankful for the solar radios. These provide a direct line for the Gospel message right into homes and courtyards of Burkinabe. "

We at Galcom are always thrilled when we hear these kinds of testimonies where the Word of God is reaching people who could not be reached in any other way. Christian radio seems to be a way that crosses the cultural and religious barriers of the Muslim world.

Mozambique
In February 2000, Mozambique was hit by a series of hurricanes that caused wide spread flooding. Now tens of thousands are homeless and starving. Jodie Nelson, with Operation Blessing, was in Mozambique after this series of disasters and was working with other relief agencies to provide food and other basic items for survival to these flood refugees. Galcom joined with Trans World Radio (TWR) to take God's message of hope to the same refugees. TWR installed a new Christian FM radio station that reached into the refugee camps. Galcom provided fix-tuned radios tuned to this TWR station to provide the Word of God to the refugees. These are not Christian people, but due to their dire situation they are very open to the message of hope in Jesus Christ. Natural disasters can open a door for the Gospel.

We received the following praise report: "Praise the Lord that the new radio station, Radio Capital in Maputo, Mozambique, officially went on the air this past Sunday. Of course, that is the station which broadcasts to the 1000 Go-Ye radios that you just produced and gave to the local church leaders to distribute there in Maputo. We believe that thousands are being blessed by the Gospel messages and as a result their lives will be turned to the Lord. Please accept our sincere appreciation and gratitude for the gift."

From Tenese Bassett of HCJB World Radio, in South Africa

"Thank you for the Galcom radios that you sent to us. We just got a response back from South Africa recently. They said that they could not pass out the radios fast enough, and that the folks there are calling the radios their blue Bibles because many are illiterate and this helps them hear the Lord's words. Praise God for His work in Africa."

Canelands, South Africa From an article in the Toronto Star, World section. (Article entitled: *Radio Doctor is on the air - Growing Audience is tuning in for HIV/AIDS advice)*

"Solar powered radios manufactured by Galcom, a company in Hamilton, Ontario are the only link many patients in squatter camps and townships in the rural areas outside Durban have with a doctor."

This was only a brief comment about the Galcom radios but again we are thrilled to know that these AIDS patients are listening to Christian radio while on their beds. They certainly need to hear the Good News of Jesus. This Christian station transmits the Gospel and Christian music and also gives them advice from their doctor on how to prevent and treat this disease.

Sierra Leone

In 2001 this country came out of an eleven year civil war. There is now a race between the Muslims and Christians for the hearts and minds of these people. A pastor in Sierra Leone, Shedonkeh Johnson, has established a strong Christian work in Bo, Sierra Leone, and has recently been granted a license for a Christian radio station in Bo with a frequency permit to transmit on FM 97.7.

In October of 2003, Galcom installed the first Christian radio station in Bo, Sierra Leone. This station is being operated by people from Pastor Shedonkeh's Church. Pastor Shedonkeh faithfully stayed and worked in the church through the eleven years of civil war, even though two of his sons were severely injured during the war.

Bo is a city of nearly one million residents but is still only one step above village living on a large scale. Central electrical power is unreliable and is often off. Battery power is a common source of power. There is no central water system. Water is hauled in buckets from wells around the city. Pastor Shedonkeh gets electric power for his church and home using generators and he is operating the radio station by generator when the central electric power is off. This Christian radio station has the capability to reach and disciple over one million people.

Pastor Shedonkeh is a great, great grandson of one of the slaves on the slave ship Amistad that landed in the US. After a lengthy legal battle and trial the slaves were set free and returned to Sierra Leone. Shedonkeh's great, great grandfather had become a Christian while going through this trial. I recommend checking out this movie "Amistad" from the video store. This is a great movie based on this true story.

The first report we received from Pastor Shedonkeh after the station in Bo went on the air was delivered to us by Jerry Trousdale of Interdev as a result of a trip he took to Sierra Leone. Here is what he told us:

"There is great excitement in Bo about the Christian radio station - FM 97.5. Trinity radio is proclaiming the Gospel of Christ on the airwaves over Sierra Leone, to this multitude who have never heard the good news of Jesus Christ. Pastor Shedonkeh says he has been overwhelmed by the response and testimonies. Letters and calls are coming in from Bo and elsewhere for prayer, music requests, etc. The station is receiving reports from listeners up to the Liberian border and from parts of Freetown, over 100 miles away, plus the neighboring towns of Makeni, Kenema and the Moyamba district. Even in the bush areas you see people walking around with a radio to their ear. This wonderful coverage far exceeds previous expectations. The community is joyful about having this medium in an society where much of community life has been pressured by growing Islamic influences.

"Pastor Shedonkeh says some Muslims are reporting that this station has become the only station that they listen to. In the first month of broadcasting many of the local churches have experienced new people attending services as a result of the broadcasts."

Uganda

The Madison Baptist Church of Madison, Alabama, has an outreach into Uganda, Africa. The pastor, Tony Starke, went to Uganda 13 years ago on a short term mission trip. While there, God spoke to his heart about moving to Uganda and starting a mission work there. In less than a year from that encounter with God he was in Uganda.

In the last 13 years he has planted 11 churches, started a Bible Institute where he has trained 43 African men to become pastors, and started a Christian radio station, New Life Radio FM is reaching a 50 mile radius in Bohembia, Uganda.

A few years ago he purchased 125 Galcom radios to distribute to remote villages in the radio broadcast range. His concept is to send a team into a remote unreached village where they stay for three weeks proclaiming the Gospel with teachings. They then leave four Galcom radios for follow-up so that the people can be discipled. After a few months they are able to recruit villagers for the Bible Institute. When these men finish Bible school they go back to their village to plant and pastor a church. He has had great success with this outreach approach using Galcom Go-Ye radios as a primary tool to disciple villagers and plant churches.

Our Lord, The Bridegroom, is calling forth His bride from all over Africa.

Chapter Fourteen

Reaching Other Nations

"And this Gospel of the Kingdom shall be preached in all the world for a witness to all nations, and then shall the end come." Matt. 24:14

According to this scripture the key indication of when the end will come is when the Gospel of the Kingdom has been preached all over the world, to all nations. Using Christian radio, and providing people with ears to hear (radios tuned to these Christian stations) the fulfillment of this scripture is possible in our day.

Galcom International has taken the Gospel already to 118 countries, giving the people in these nations ears to hear. The following are some additional testimonies of the Gospel reaching into other nations by Go-Ye radios, giving them ears to hear.

Istanbul, Turkey
A few years ago, at the request of the Southern Baptist Mission Board, Galcom International built a one watt radio transmitter in a suitcase to be used as a mobile unit in various locations in Istanbul by a missions group there. This transmitter reaches out to about a three mile radius and has been used to broadcast the Gospel to some of the 7,200,000 Islamic people in Istanbul.

Since that day Turkey has instituted new licensing requirements, allowing all agencies who had been previously broadcasting to be grandfathered in and given a full broadcasting official radio license. Now this mission group has a major FM radio station in Istanbul that is proclaiming the Gospel to the whole city. Our little Galcom radio transmitter in a suitcase opened the way to reach this whole Islamic city with the Gospel. You never know how God will use something that has such a small origin to magnify His Holy Name.

The Philippines, Resources for the Blind

Over a four year period we have sent several thousand radios to Resources for the Blind in the Philippines. Their director, Randy Weisser, has sent us the following report:

"Greetings from the Philippines! I would like to introduce to you Pastor Jose Ibaanez. He is one of our blind staff that is instrumental in distributing the Go-Ye radios. His assignment is to travel throughout the rural regions of central Luzon, finding those people who are blind like him, and conducting Bible studies in their homes. After his Bible studies, he leaves one of the Galcom radios with the blind person.

"As you probably know, most of these blind people live very lonely, isolated lives. In most cases, they seldom venture out of their homes and are thus really an unreached people group. There are an estimated 500,000 such blind persons in the Philippines. Recently I asked Pastor Ibanez to carry a small recorder with him and try to get feedback from some of his clients about the Go-Ye radio. He received many testimonies.

"Thank you so much for helping us to reach this very receptive, but difficult to locate group of people. I am sure that these people listen to the radio more than any other group of people. We just met one of the pastor's clients who wears the radio around his neck. At night, he ties it to his wrist so that no one will be able to take advantage of his blindness and steal the radio from him. Here is just one of many testimonies of the blind being blessed by the Galcom radios distributed by Pastor Ibanez."

Estonia

Pastor Endel Meuisi from Estonia reported that he had gone into a local prison where 650 of the most violent prisoners were held. We have provided him with Galcom solar powered radios tuned to Christian radio stations in his area. It occurred to him that many of the hard-liners housed in the nearby prison might profit from the use of the radios. He asked the warden of the prison if he could give some of the prisoners the Galcom radios fix-tuned to local Christian radio. Initially the warden was not very receptive since he wanted to avoid anything that would give rise to unnecessary disruption in the prison. Finally the warden allowed him to pass out 120 Go-Ye radios.

A few weeks later he received a call from the warden who was ecstatic. The whole atmosphere of the prison had changed since the radios were distributed. However, one problem arose that the warden did not know how to deal with. He reported, "There are 262 prisoners who want to be dipped in water and I don't know what to do." Pastor Meuisi decided to investigate. All of these prisoners had accepted Christ as their personal Savior and wanted to be baptized. The warden also asked him for more radios. Pastor Endel made arrangements to carry out the baptism and the newly formed prison church continued to have a great impact on the rest of the prison. The radios continue to proclaim the Gospel in this prison setting many prisoners free from a life of sin.

Report from Romania

Allan McGuirl, Galcom Canada, was talking to Richard Ellison, who is with Helps Ministry International in Romania, and he is so excited about how God is using the Galcom fix-tuned radios in various Gypsy villages in Romania. The people treasure these radios. In many places Gypsies are a rejected people, but Richard has targeted his radio programs to reach them with the Gospel.

One blind lady who was saved runs a massage parlor and has her radio going all the time. The Lord has used her to get radios into poor, needy areas.

Richard advised Allan that the license had just come through to increase their power from 100 watts to 1,000 watts. He said that the people there call their station "SOS Radio 101.4". Allan asked, "What does the SOS stand for?" Richard said, "Save our souls." We think of it as a distress signal and they refer to it as a call signal to salvation.

Richard also said that a doctor has his little Go-Ye radio in his operating room while he does surgery. Everyone in the room has an opportunity to hear the Gospel daily. He went on to say that a powerful secular station in Bucharest, which is 40 kilometers away, has the same signal and the Lord intervened and they changed the frequency of the secular station. God answers prayer. We thank the Lord for stories like this.

Micronesia

Allan McGuirl, the International director of Galcom in Canada, went to Micronesia with a team and installed a new 500 watt Christian AM radio station. No one is really sure where the people of Micronesia came from. Micronesia at one time was called the North Caroline Islands. Since WWII they have divided into separate nations and states. Throughout this area of the Pacific, that is north of the equator between Hawaii and the Philippines, there are the nations of Palau, Republic of the Marshalls and the Federated States of Micronesia. In addition, there is the territory of Guam and the commonwealth of the Northern Marinas. The Federated States of Micronesia (FSM) is made up of the four states of; Korsae, Pohnpei, Yap and Chuuk.

This station is reaching 45,000 people living on 98 tiny islands in the state of Chuuk where only a very small percent of the people are Evangelical Christians. The people are very resistant to the Gospel. This is where radio has an open door. While they will not come to church or even listen to you, they will listen to the radio. God's Word and God's Spirit, in time, can break down the resistance. With this Christian AM station and our solar powered radios we will be reaching thousands of people right in their homes throughout these 98 small islands.

Middle East

Don McLaughlin, of High Adventure Canada, was in Israel in early 2003 going through some old files from the radio station there. He came across five or six containers, stacked with cartons containing thousands of letters from listeners. As Don started to read them, he was stunned with the awesome testimonies of the listeners. There were about 2600 letters in each carton. We just thank God for His strategic timing in these remarkable radio harvest testimonies.

Don reported this to George Otis and in turn George sent the following letter to the Galcom office, which I feel is worthy of including in its entirety:

"Dear Friend,
I was sitting in a small hotel in Northern Israel, near the border of Lebanon and Israel, when a distinguished looking man walked up to shake my hand. He had come to share a startling report from his completed journey through Syria, Lebanon and Jordan. He is the Bishop of that territory from the Maronite Church.

"He told how dozens of small churches were springing up in these heavily Muslim nations because of the day and night ministry over the Holy Land Christian radio transmitters. He spoke of salvations, healings and other miracles through the hearing of God's Word over the airwaves. He said, 'Please don't stop!'

"But you know, it took two provisions to light up those skies in such a hostile region. The first is the Christian station broadcasting the glory of Jesus. The second requirement, and all important, the radio receivers for these truths to be heard. The Bishop described how five to twenty hungry souls would often be sitting on the floor listening around one small radio.

"A year or more earlier Harold Kent, a pioneering visionary believer from Florida, had sent us the first of some 70,000 remarkable Galcom sun powered radios to distribute, at no charge, to the people in those three nations, plus Israel.

"The Bishop's testimony illuminated God's vision for reaching the unsaved. The Lord said to him, 'He who has ears to hear, let him hear.' Through these radios, that Word was being fulfilled. Galcom, clearly has come to the Kingdom for such a time as this! The radios can also be found in Iraq, Iran and Saudi Arabia."

George Otis added the following to this letter:
"Nearly 420,000 of these Galcom radios are now scattered on portions of every continent. But we must remember this - there is an urgent and almost desperate call to reach into those places which have no way of hearing His Word, God's call to salvation.

"The Lord will greatly bless each of us if we will pray intensively for Galcom to be able to speed up its distribution of these proven tools for this, our Lord's final harvest of the lost. Extremely important will be our financial support for this difficult, but vital ministry, to allow the rest of the unsaved to hear God's call while there is yet time. Please send the largest gift you possibly can this week.
May Jesus bless you greatly,   George Otis"

Chapter Fifteen

Memories of Travels to Many Lands
By: Allan McGuirl

"He that goeth forth and weepeth, bearing precious seed, shall doubtless come again rejoicing bringing his sheaves with him." **Psalm 126:6**

It has been interesting to see and be a part of how God allowed each one of us to have a part in the ministry of Galcom International. Ken Crowell, has a heart to bless Israel by producing the first radios in Israel providing employment, and also has deep concern for outreach to the millions on the mission field who had never heard the Gospel. God has used Harold Kent, who along with his dear wife JoAnn, has sponsored literally hundreds of thousands of radios around the world. Gary Nelson, head of the Galcom USA ministry, has been kept busy selecting the most vital locations for radios along with raising funds for the ministry. In the midst of this, God has allowed me to make contact with various ministries around the world and in many cases travel to many locations to either plant radio stations, to distribute radios or to investigate new opportunities for outreach. The following stories are precious memories of my many mission trips over the years and how God moved in each situation:

Haiti: Installation of the first low-powered radio station

Ross Robins, the director of Light of the World Ministries, was ministering in Haiti. Early in 1993, he spoke to me about their desire to set up a Christian radio station in the southern part of Port au Prince to reach out to many of the poorer people who had no radios at all and needed a Christian witness in their community. He had already conferred with Pastor Simone who had a Christian school, small orphanage, and a church in the area and was willing to be involved in radio ministry. Funding was sparse for work among the poor so he had no money to set up a station, put in a studio or pay for the gas to run a generator. It was determined that he would need a 50 Watt AM station. A friend of Galcom, James Cunningham, had just designed a 50 Watt transmitter for this type of application.

A few weeks later, while sharing the Galcom ministry at a small Presbyterian ladies meeting on a Wednesday evening, I closed by saying, "By the way, here is a prayer request. We have a pastor in Haiti who wants to set up a Christian radio station in the southern part of Port au Prince. Is there any possibility that you can pray that God would provide the money?" "How much would that require?" was the next question. I responded, "Well, we need about $2,500 to do the station." No other comment was made.

A little later, while having light refreshments, one of the ladies requested that I drop by her home the following day to pick up a little gift for the ministry. Much to my surprise, she presented me with a check for $2,500 to install this station.

In June, 1993, James Cunningham and I left for Haiti to install our very first low powered Christian radio station. The building for the station was in a crammed slum section of Port au Prince surrounded by a high wall imbedded with broken glass to deter thieves. An AM antenna requires enough land to lay out a grid of ground wire. Pastor Simone had some land but not nearly enough for the antenna grounding. Then the Lord gave James an idea, "Let's hook the antenna up to the water pipe coming into the building as a part of the ground." We found the water pipe and made a solid connection. Praise God, we were able to complete that. Then we had to put in eight foot grounding rods, but there was a cement wall poured all around the building. A nearby missionary loaned us a cement drill and we were able to complete the grounding job. By God's grace we persevered in 104° heat.

    Doing the tower work was interesting. The mission had a 40 foot rickety, old, wooden ladder to reach the top of a tower to make the antenna connections. Putting the ladder on the roof of a building gave us additional height even though it leaned precariously over the wall with the broken glass. As I ascended up over this wall, suddenly a creaking sound jolted me into the realization that the ladder was disintegrating under me. "Lord, I need your help!" It stayed intact just long enough for me to clear the broken shards of glass and land on the roof, a little shaken but unharmed.

    Hours later we had our dear brother turn on the switch for the transmitter and a beautiful sound of Gospel music came out. What a joy! Everyone was thrilled and this community now has a new radio station broadcasting the Gospel to needy souls.

    Since the air was very hot at night, we had been staying in a mission compound up on the mountainside. The gas embargo was on and very few stations had fuel. As we descended this mountain road to leave on Saturday morning, we came around a bend only to find it blocked by cars and trucks trying to get into a gas station. Just then a Haitian military truck came roaring down the mountain and pulled in front of us. Then the military leader called his men to go up front to clear the road so they could get through. Ross then knew it was safe to follow reasonably close behind the military vehicle. They had cleared the way allowing us to get to the airport on time. Praise God.

This Christian radio station has been used so effectively over the many years. Although Pastor Simone has gone on to glory and his son has taken over his work, the legacy of that Christian radio station continues to be felt in very real ways. Praise the Lord.

This station was our first station and was set up to run with back up power coming from either solar panels or a gas generator, as there are many power interruptions in Haiti. All of this happened because one lady obeyed God and gave a gift of $2,500. Only eternity will tell the entire story and the end results.

Poland: Another station installation
As a delegate at the National Religious Broadcasters Conference (NRB) in 1996 in Washington, I was looking for a place in the 6,000 seat auditorium for one of the evening sessions. Finding a single seat, I began to converse with the lady next to me. Eva Brycko was from Warsaw, Poland, and had come to the NRB Conference to see if she could talk to somebody about setting up a Christian radio station in the town of Ostróda, about

a three hour drive north of Warsaw, where 150,000 people lived. There was no Christian outreach in Ostroda other than a couple of small churches. In that huge NRB convention God had placed me right beside the person that needed to know about low powered radio broadcasting!

This meeting began several months of correspondence, prayer and planning. The next spring I was at the Peoples Church in Toronto and I shared about the need for a radio station in Ostróda. They provided a good portion of the money for the transmitter and antenna. Later, while sharing at the Peoples Church in Montreal, they agreed to provided the balance of the money needed for a studio.

In mid-October, my son, Allan David, and I went to Ostróda to install the station. About six believers from this town had previously bought a used 100 year old stone water tower about 100 feet high which was situated on the highest point of land in the town. It had been bombed during WWII and was no longer usable for water storage. Apparently, the bomb had hit right in the center of the tower, smashing through the roof. The large volume of water had absorbed the impact so that it didn't explode, it just continued on down through the floor of the reservoir and landed at ground level. When they entered the tower the bomb was still intact requiring removal by the military.

As Allan David and I ascended the rickety old stairs, we could see the hole in the floor where the bomb had gone through. The tower provided an excellent location for the antenna. We also installed the transmitter and connected it to a nearby building where we had set up the studio equipment.

I will never forget the thrill and excitement of everyone there as we flipped on the main switch in the studio and began transmitting. We turned on some of the Galcom fix-tuned radios that we had brought with us and all of a sudden everyone began to shout; "WOW, it's on the air, it's working."

Today, Radio Mazury in Ostroda, Poland, is still broadcasting and many thousands of people continue to hear the Word of God. Many radios have been given out and many lives have been touched in a very special way. Continue to pray for Radio Mazury as they broadcast the Word of God in that needy area.

PERU

Another memorable trip was when I met Dan Muth in Cuzco, Peru. Dan needed advice on a new radio station project to reach the 100,000 Inca Indians living in the Sacred Valley of the Incas near Machu Picchu. I had brought solar powered radios fix-tuned to an existing Christian station to distribute in the area. We left Cuzco in a four wheel drive vehicle and started climbing and climbing. As we reached Machu Picchu at 14,000 feet my head began to feel light and I felt as though things were moving in slow motion. However, it was encouraging to see the ministry the Muths have among these people.

On Sunday I was asked to preach, while Dan translated into Spanish and an Inca interpreter relayed the message in their mother tongue. Consequently, the sermon took a lot longer than normal. I was humbled to gather in this little thatched roof church holding about 60 people. The Inca pastor proceeded to celebrate communion. What a precious time! The bread was like little rosettes twisted into a beautiful design and served in the lid of an empty chocolate box. The juice was in about 30 cups in the bottom of the box. When they had served us they refilled the cups and served the remainder of the people. Sparse resources perhaps, but their praise to God and their soft singing allowed me to experience one of the most beautiful communion services I have ever participated in.

Following the communion service, they provided lunch. Dan cautioned me that they would be giving me something special. Soon, an older lady came in with a tray. There was corn, potatoes and a completely skinned and roasted guinea pig. The custom required me to share this delicacy so I started on a hind leg. It tasted something like dark turkey meat. However, the head and some of the other parts did not appeal to me. I broke off the head and was inclined to toss it to a dog nearby. Dan, sensing my thoughts, suggested I give it to one of the Incas. I gladly offered it to a big fellow nearby who gladly accepted it and, sitting across from me, proceeded to crunch his way through the skull and entire head. To the Incas this was a delicacy.

I left that country rejoicing at the faithfulness of the Lord and the missionaries as they minister to these Incas. I was praising the Lord for the opportunity to prepare another radio station to reach into these mountains and valleys. Subsequently, we have shipped several thousand more Galcom Go-Ye radios for these people.

Lebanon
The very first fix-tuned radios produced in Tiberias, Israel, were being tuned to Voice of Hope radio station in the Middle East. Arrangements to ship them to Lebanon had been made from which

point some would find their way into Syria, Jordan and, as we learned much later, into almost every Middle Eastern country.

One colonel in a UN camp in Lebanon received one of the little radios and began to listen. The first radios had no speaker, only a set of small ear buds. At this camp twenty two other men crowded around to hear what they could of the messages from this one radio. The colonel wrote our office in Tiberias where Ken Crowell was building the radios, requesting twenty two more radios for the men at the camp as quickly as possible because all these men wanted to hear the Word of God.

Belize

Early in the Galcom history we installed a radio station in Belize. I first met Richard Smith from Belize at an NRB Conference. He was young but very interested in setting up a Christian radio station in his country; however, funds and experience were very limited. As we met over lunch, I encouraged him to follow the Lord's leading and leave the results with Him.

Richard commenced contact with the department of communications, secured a station site and was assigned a radio frequency. Gradually, support materialized to purchase the equipment and set up the radio station. Now, five years later, the Lord has enabled him to set up several other stations besides the mother station. He has helped several others get started in Christian radio as well saying, "I would not have done this if Galcom had not challenged me to step out on faith and trust the Lord to do it." We give God the glory for using Richard in an effective way to reach his people in the country of Belize.

Guatemala

I remember making contact with Pastor Fausto Cebiera and his wife Miriam. They have a very precious radio ministry in Guatemala using a number of radio stations with a burden to reach into the northwest of Guatemala to the Ixil Indians.

On one trip I was in Guatemala for a COICOM Conference and took advantage of the opportunity to go with the Cebieras into Ixil country. We drove several hours by jeep through windy roads into the back country. We were heading to the town of Nebat where Pastor Cebeira had an FM station. On one part of the road it took us three hours to go 17 miles because of road conditions. Along one stretch half of the road was washed out by a roaring river.

Along the way we handed out Galcom Go-Ye radios to the Ixil people we met. I still recall one lady struggling down the trail with a huge pile of wood stacked on her back. We stopped to offer her a radio. The sparkle that came to her eyes and the smile on her face were indescribable as she turned on the radio and heard the Gospel in her own mother tongue. As she continued on down the trail I prayed that she would soon know the Savior.

Pastor Cebeira has touched the lives of thousands of people throughout that whole area and is doing his very best in setting up low powered radio stations and distributing Galcom fix-tuned radios.

I had the privilege of preaching and eating with Ixil Indians and sensed their love for Jesus. George Calder, a retired friend from Hamilton, accompanied me on this trip. He was just thrilled to see how God was using radio to reach these people.

As darkness began to settle in, Miriam was getting concerned about the need to get back to Nebat to the mission house. Fausto, our driver, was experiencing difficulty as it had started to rain very hard and the road was becoming slippery and was filled with ruts. Just a short distance down the road the jeep went off the road into the ditch. George, Miriam and I got out and tried to push and shove but to no avail. Out in the pouring rain, surrounded by the misty cold of the mountains, it seemed like we might have to spend the night there. I will never forget Miriam standing out on the road in the pouring rain asking God for help.

Her big concern was the guerrilla activity in this part of the country. Many expatriates had been kidnapped for ransom and her concern was for our well being. Within moments of her prayer we heard all kinds of yelling and howls. Down from the hill came twenty-one Ixil Indians. They soon got out the pails, jacked up the vehicle and lying down in the mud placed rocks strategically underneath the jeep. They were able to move us ahead a short distance. They repeated this procedure until we were able to navigate the trail once again. God had answered prayer instantly. Arriving back at Nebat cold and tired we were grateful for a good meal and a bed to sleep in. The Cebeiras, who are now in their mid-sixties, have a great zeal for the lost Ixil people.

There has been much growth among the Ixil people of Guatemala. Churches are growing, leadership is being developed and new believers are being discipled in the Word. This is made possible through the little Galcom fix-tuned radios. Faust and Miriam Cebeira cannot travel often into these difficult mountain areas, but with the radio ministry, they can reach all of the people throughout the area every day. The results are mushrooming.

In all of these accounts, it is evident that God has been using His vast network of believers to accomplish the work of His kingdom. We, the Galcom team, have been privileged to be a part of this great network. From national believers to missionary workers to volunteers, God has enlisted all of us to do His bidding.

The work of Galcom International continues and will continue until all nations are reached for His Kingdom and our Lord and Savior returns for His Bride. At that time, those of us who have labored in His rich harvest field will come rejoicing bringing in the sheaves.

Chapter Sixteen

The Stones Are Crying Out Plus a 2009 Radio Update

"And He answered and said to them, I tell you that, if these should hold their peace, the stones would immediately cry out" Luke 19:40

As Jesus rode into Jerusalem on a donkey He made this statement in Luke 19 that someday the stones would cry out with the Word of the Lord. That is now happening with the MegaVoice audio Bibles where we are storing the Word of God on a silicon chip (compressed sand, a man made stone) and sending them all over the world. Even Joshua talked about God's Word being stored on a stone (Joshua 24:25-26). The MegaVoice is a solar powered audio Bible. With most of the world's population being oral learners we are now able to provide them with the Scripture in a simple audio microchip player that is solar powered.

Many people around the world are coming to a saving knowledge of Jesus Christ through Christian radio. However, the Great Commission says to make disciples, not just converts. Disciples must have the Word of God in their own language. We now have the MegaVoice as a tool to provide the Word of God to the non-literate in their heart language.

One of these Scripture outreaches that Galcom USA is very involved with us to the many native Indian tribes in the Southern Mexico state of Oaxaca.

Jim and Jamie Loker, of Missionary Ventures, are working in Oaxaca recording the New Testament in 35 different languages of the tribal people there. Years ago Wycliffe translated the Bible into these languages but most of the people are non-literate, they cannot read. Now they are finally receiving the Bible in their own language in audio form loaded on solar-powered MegaVoice microchip players. We have supplied them with 3100 MegaVoice players so far. With MegaVoice they will have the ability to provide audio Scriptures to 465,000 native people. Here is a recent report from Jim's wife, Jamie Loker:

Reaching People of Oaxaca with Megavoice

God wants His Word communicated to the 120 people groups of Oaxaca, Mexico. Sound simple? It isn't. Although many languages have a translated New Testament, less than 1% of the indigenous population can read. Christian radio is nonexistent.

Does this scenario sound hopeless? Maybe to some it sounds overwhelming at best, but we have found a solution. Just over a year ago, Galcom provided our ministry in Oaxaca with the first 2500 MegaVoice players. These have been distributed among **twenty different language** areas.

Jim and I began ministering in southern Mexico in 1994. Since 2002, Jim and his team have recorded more than 35 complete New Testaments. The goal is to see God's Word made accessible to all the people groups of Oaxaca, and eventually to all of Mexico.

One of the native Pastors, Pastor Juan, recently told us, "Brother, we don't have electricity in our village." This is a problem in many rural areas of

Oaxaca which limits the effectiveness of Scriptures on cassette or CD. Pastor Juan is one of many who are thrilled to learn that the MegaVoice players do not require electricity. Instead they utilize solar panels and rechargeable batteries. This was an answer to prayer for Juan's village. His people can now listen to God's Word in Chinantec, their native language. As people listened to Scripture they commented, "Others have tried to deceive us with their words and beliefs. Now we are hearing God's Word and believe it. We know it is the truth".

Another group, the Amuzgo people first received their written New Testament in 1980. Few are able to read, but many are hungry to hear God's Word. MegaVoice players have been distributed to more than 30 missions among the Amuzgos. For the very first time, they are hearing God's Word in their own language. Among those listening to the Amuzgo Scriptures are many elderly folks with failing eyesight. It may be too late to teach them to read, but they can hear God's Word on MegaVoice players. To date, we have distributed MegaVoice New Testaments to more than 35 families with elderly members, including one who is in his 80's, another who is 90, and one man who is 104 years old.

A native pastor from another tribal group recently called us and I couldn't believe what he was telling me. The first 50 MegaVoice players had just been distributed among the people of San Lorenzo, and he was begging me for 150 more units. He reported that there are seven villages that speak his language. Most of the people cannot read, but want to hear God's Word. We gladly provided them with more players.

The Chatino people are farmers. Leaving before sunrise, they work all day, not returning home until late in the day. With the introduction of the MegaVoice players, many are taking advantage of their time outdoors by listening to the Scriptures in Chatino. They carry their players in their shirt pockets, and listen while they walk or work. When the battery needs to be recharged again, they put it in the sun for a few hours, and start listening again.

Is reading a requirement for people to learn God's Word? Certainly not, or else 2/3 of the world's population would be disqualified from becoming followers of Christ. Thankfully, the Scriptures assure us that "Faith comes by *hearing* the Word of God." There are people all over the world who cannot read and will never learn to read. Praise God with us that MegaVoice is filling this void and the rocks are indeed crying out with God's word. **Isa. 2:3b says "For out of Zion shall go forth the law, and the word of the LORD from Jerusalem." The MegaVoice players are produced in Israel and sent out from Zion into all the World, a prophecy fulfilled.**

A 2009 Galcom Radio Update

Since this book was first published in 2004 hundreds of thousands more Go-Ye radios have been produced and distributed around the world into 126 countries. As of the end of 2008 the number of these tiny radio missionaries sent out has reached over 730,000 plus 90 radio stations have been installed.

One Christian radio station installed by a Galcom donor foundation in West Africa reported 60,000 people who have come to know Christ because of radio and the ministry of the local church that hosts the radio station. In addition they have planted 975 new house churches as a result of this great harvest of souls. The radio station combined with the Galcom Go-Ye radios are serving as great discipling tools as well as for evangelism.

There are so many more stories of how God is using Galcom and these little radio missionaries that we could fill several more books. Some stories can be read in monthly newsletters posted on www.Galcomusa.com and www.Galcom.org. **They are stories of what God is doing** though generous donors and the Galcom International team in both the USA and Canada.

Acknowledgement of donors

This book and the testimonies contained herein are only possible through the generous support of our donor partners now numbering over 500. But it is really by the grace of God that His Word in going out to millions of listeners in 118 countries as the Go-Ye radios give them <u>ears to hear</u> the Good News.

We want to thank all of our partners who have obeyed the Great Commission by supporting the work of Galcom International as we labor in the harvest fields of our Lord.

Some, such as Herb and Lola and Don and Dorothy have sold houses and generously tithed on the sale to this work. Others have supported projects through Trusts and Foundations. The Festus and Helen Stacy Foundation has funded thousands of radios and are now funding radio station projects with the first station operating successfully in Sierra Leone.

Harold and JoAnn Kent and The Kent Family Foundation funded the start-up of this ministry and have sustained the work of Galcom generously beyond imagine for the last 15 years. Without their contributions Galcom International would not exist.

There are so many faithful saints who give each month so that God's work can go on. A group of ladies from the Trout Lake Bible Study put together a cookbook to raise money for Go-Ye radios.

It is impossible to thank everyone individually, but be assured each and everyone is appreciated by all of us in Galcom and they are also laying up treasure in heaven. Only what is done for the Kingdom of God has any eternal value. I will relate just one story of a widow who has a small income but a big heart and a bigger vision for missions.

Widow's Mite:
Pennies multiply for radios: A long time supporter of Galcom, Marie Eiler of Tampa, Florida, saves pennies to donate to Galcom for Go-Ye radios. She is a widow on a small pension and also takes in ironing to help raise her support. When shopping, she asks for her change in pennies. She always carries a Galcom radio in her purse to show the clerks what she uses the pennies to purchase. Now several of her friends are also saving pennies for her. The ladies in her daughter's church group are also saving pennies for her. The pennies keep multiplying and she is able to send more and more little talking missionaries, Go-Ye radios, out onto the mission field, proclaiming the Good News of Jesus Christ to the lost.

She told us: "When I was a child we were very poor and got evicted from our home. While we were on the street a man came by and gave us children each a penny so we could go to the store and buy candy. Ever since that day a penny has always been very important to me." In the last six years she has donated over $400 in pennies, weighting over 250 pounds. This is enough to send 25 radios to the unreached. Hundreds of people are being reached with these 25 radios, as they listen to the gospel daily. She has given them <u>ears to hear</u>. She is a modern day <u>hero of the faith</u>.
Thank you Marie

To find out more information about the ministry of Galcom International or to become involved as a partner in this Great Commission work contact:

Galcom International USA, Inc.
P.O. Box 270956
Tampa, FL 33688-0956
Ph. #(813) 933-8111
Email: Galcomusa@galcom.org

Or in Canada

Galcom International CN
115 Nebo Road
Hamilton, Ontario L8W 2E1
Ph. # (905) 574-4626
FAX # (905) 574-4633
Email: Galcom@galcom.org

Visit our web site at:
www.galcomusa.com
or
www.galcom.org